Modern Sound Reproduction

Modern Sound Reproduction

HARRY F. OLSON, E.E., PH.D., D.SC. HON.

Consultant and Staff Vice President Ret.
RCA LABORATORIES
Princeton, New Jersey

VNR VAN NOSTRAND REINHOLD COMPANY
New York / Cincinnati / Toronto / London / Melbourne

Van Nostrand Reinhold Company Regional Offices:
New York   Cincinnati   Chicago   Millbrae   Dallas
Van Nostrand Reinhold Company International Offices:
London   Toronto   Melbourne

Manufactured in the United States of America

Published by Van Nostrand Reinhold Company
450 West 33rd Street, New York, N.Y. 10001

Published simultaneously in Canada by Van Nostrand
Reinhold Ltd.

15   14   13   12   11   10   9   8   7   6   5   4   3   2   1

Library of Congress Cataloging in Publication Data
Olson, Harry Ferdinand, 1901–
   Modern sound reproduction.

   Includes bibliographies.
      1. Sound—Recording and reproducing.   2. Electro-acoustics   I. Title.
TK7881.4.045   621.389'33   71-185982

Preface

Sound-reproducing systems are exemplified by the telephone, the phonograph, radio, the sound motion picture, magnetic tape, television, hearing aid and re-enforcement. Universally employed in all variations of modern living, these systems constitute indispensable media for the dissemination of information, art, and culture.

The main purpose of modern sound reproduction is to provide the listener with the highest possible artistic and subjective resemblance to the live condition, or with suitable modifications that will improve the intelligibility or increase the artistic or emotional impact of the sound.

Although sound reproduction has been in existence since the turn of the century, there have been continuous improvements in performance over this period. Outstanding and significant developments in elements and new systems have been made in the past decade. Accordingly, this is an especially opportune time to provide a book on the specific subject of modern sound reproduction.

Particular efforts have been directed toward writing a book that will provide useful information to a wide range of readers, including scientists, engineers, technicians, and audio laymen and enthusiasts. As an aid in attaining this objective, the book is illustrated with 250 figures. In many instances the figures contain several parts, so that an entire theme is presented in a single illustration. To provide further assistance in interpretation, the illustrations are delineated to depict the physical action and performance. The major portion of the book employs simple physical explanations and descriptions which can be read and understood without any special training in engineering, physics, or mathematics. However, for the trained engineer and scientist, the book is backed up by technical descriptions of the action, performance, and characteristics of sound-reproducing systems, including the use of dynamical analogies.

The elements employed in modern sound reproduction—microphones, amplifiers, loudspeakers, and earphones, and the systems of magnetic tape, radio, phonograph, sound motion picture, television and sound-reenforcement—are described in considerable detail. Included are the generic monaural, monophonic, binaural, stereophonic, and quadraphonic sound-reproducing systems. Room acoustics are applied to describe the performance of studios, theaters, auditoriums, and rooms and the relation to sound reproduction. The fundamental acoustical measurements that play an important part in the advancement of sound reproduction are described. The ultimate destination of all useful and informative reproduced sound is the human ear. Therefore, also included are the characteristics depicting the action of the human hearing mechanism with relation to sound reproduction.

A few words about the subject matter in relation to other books by the author may be of interest to the reader. In the book *Solutions of Engineering Problems by Dynamical Analogies* the objective is to provide means for solving problems in vibrating systems by a reduction to the analogous electric circuit. The book *Acoustical Engineering* is devoted to the subject of engineering acoustics in all phases and aspects and includes the fundamentals of acoustics, the theory, design, and measurements of electroacoustic transducers and systems used in sound reproduction, ultrasonics and underwater sound, and applications in these fields. In the book *Music, Physics and Engineering* the objective is to present descriptions and expositions of the processes, instruments, and characteristics involved in all the steps from the musical notation on paper to the ultimate useful destination of all sound, the human hearing mechanism. And, as outlined above, the main objective of this book is to present a detailed technical exposition on the significant and essential elements and systems of modern sound reproduction for a wide range of readers including scientists, engineers, technicians, and audio laymen and enthusiasts.

HARRY F. OLSON

Contents

6   SOUND-REPRODUCING SYSTEMS   124

7   TELEPHONE, HEARING AID, AND SOUND SYSTEMS   134

8   MAGNETIC TAPE RECORDERS AND REPRODUCERS   142

13   ACOUSTICAL MEASUREMENTS   229

14   ACOUSTICAL PERFORMANCE OF ENCLOSURES   249

15   SOUND COLLECTION IN STUDIOS   262

16 SOUND DISPERSION IN ROOMS, AUDITORIUMS, AND THEATERS 280

17 SUBJECTIVE ACOUSTICS 304

Modern Sound Reproduction

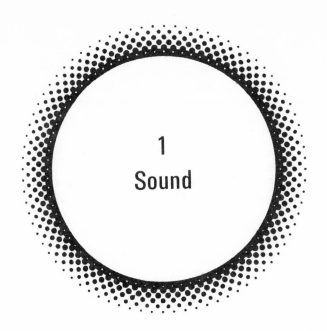

1
Sound

1.1 INTRODUCTION

The reproduction of sound is the process of picking up sound at one location and reproducing it either at the same location or at some other location, either at the same time or at some other time. The most common sound reproducing systems are the telephone, the phonograph, the magnetic tape reproducer, the radio, sound-reenforcing systems, sound motion pictures, television, and television reproducers.

The telephone is the oldest sound-reproducing system. There is, on the average, more than one telephone instrument per family in the United States. The telephone instrument combined with complex office and interconnecting equipment makes it possible for any person to talk to any other person anywhere in the United States in a matter of seconds.

The phonograph was the first sound-reproducing system that made it possible for all the people of the world to hear statesmen, orators, actors, orchestras, and bands; previously, only a relatively small number of people could hear them firsthand. The phonograph is used all over the world. There is an average of at least one phonograph per family in the United States. For the past decade, the sale of records has averaged five per year for every man, woman, and child. The popularity of the phonograph is due to the fact that an individual can select any type of information or entertainment and reproduce it whenever he so desires.

The magnetic tape recorder and reproducer augments the phonograph system in providing a simple recording means as well as prerecorded material for reproduction in the manner of the phonograph. The magnetic tape reproducer in the automobile has made it possible to select and reproduce any desired material in that environment.

The radio, like the phonograph, is a consumer-type instrument. Practically every family in the United States owns several radio receivers, from the small pocket personal type to the console high-quality type. More than three quarters of the automobiles are equipped with radio receivers. There are more than 6000 broadcasting stations. The net result is that nearly every person can select almost any desired program for listening.

Sound-reenforcing systems have improved the hearing conditions in large auditoriums and halls and have made it possible for a speaker or musical aggregation to provide good sound coverage for an outdoor gathering of almost any size.

The addition of sound to motion pictures made the latter type of expression complete. The system was the first in which picture and sound were synchronized and reproduced at the same time. Nearly half of the population in the United States attends a sound motion picture theater once a week.

Television is the latest broadcast system in which picture and sound are telecasted and reproduced in the home. Sound is, of course, important to television because without it the result would be the same as a silent motion picture. On the average, every family in the United States owns a television receiver.

Television recorders and reproducers of various types are now becoming available to the consumer. Prerecorded materials similar to phonograph records are now being produced. A customer can select any prerecorded material and obtain the reproduction of picture and sound when he so desires.

The telephone and the interconnecting system have made it possible for any person to talk to almost any other person in the world. The radio, the phonograph, the magnetic tape reproducer, sound-reenforcing systems, sound motion pictures, television, and television reproducers have made it possible for all of the people of the world to hear famous statesmen, artists, actors, and musical aggregations, while only a relatively small number had been able to hear them firsthand. The reproduction of sound has produced in a relatively short time a great change in the industrial complex, education, news, and entertainment of this and other countries. The impact of the sound-reproducing systems upon the dissemination of information, art, and culture has been tremendous. The reproduction of sound as exemplified by these electronic media has been as important to the advancement of knowledge as the invention and application of the printing press.

The purpose of this text is to provide an exposition and description of the latest and most important elements and systems employed in the modern reproduction of sound where the primary objective is to provide a high order of excellence of performance. However, before the specific subjects involved in sound reproduction are discussed, some of the fundamentals of sound that are related to the reproduction of sound will be considered. Accordingly, the purpose of this chapter is to provide a brief presentation of some of the aspects of sound directly involved in the reproduction of sound.

1.2 GENERATION OF SOUND

Sound is an alteration in pressure, particle displacement, or particle velocity which is propagated in an elastic medium, or the superposition of such propagated alterations.

Sound is also the auditory sensation produced through the ear by the alterations described above.

The ultimate destination of all desired sound whether original or reproduced is the human ear. Therefore, the considerations of sound reproduction in this book will be confined to the audio frequency and amplitude ranges.

The medium for the transmission of original and reproduced sound in this text will be confined to air.

The definition of sound indicates that sound is produced when air is set into motion by any means whatsoever. There are five main means for the generation of sound waves in air, namely, the vibrating body, the throttled airstream, an explosion, a thermal process, and Aeolian operation. These means for the production of sound will be described as follows.

The vibrating body depicted in Fig. 1.1A is the simplest means for the production of a sound wave. The motion of the vibrating body produces a sound wave in air which continues as long as the motion of the body continues. Some examples of a vibrating body are the sounding board of the

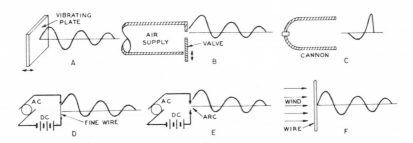

Fig. 1.1 Sound generators. A. Vibrating body. B. Throttled airstream. C. Explosion. D. Thermal. E. Arc. F. Aeolean.

piano, the body of instruments in the violin family, the guitar, the banjo, the stretched membranes of drums, cymbals, and the diaphragms of loudspeakers.

The throttled airstream depicted in Fig. 1.1B is another common means for the production of a sound wave. The throttled-airstream generator consists of a valve and source of air under pressure. The valve controls the opening of the aperture. The volume of air that is emitted is a function of the aperture opening and the supply pressure. The operation depicted in Fig. 1.1B converts a steady airstream into a pulsating one. The frequency of the sound wave corresponds to the vibrating frequency of the valve. Examples of the throttled-airstream generator are the human voice, the trumpet and other lip-modulated instruments, the clarinet, the saxophone, organ pipes, the organ and other mechanical and air-reed instruments, and the siren.

An explosion-type generator is depicted in Fig. 1.1C in the form of a short cannon. The sound wave emitted by an explosion consists of a sharp, high-amplitude compression wave of short duration followed by a lower-amplitude rarefaction wave of longer duration. Other examples of explosion-type generators are bursting balloons and tanks or enclosures of compressed gas, and so on.

An example of a thermal sound generator consisting of a fine wire connected to the combination of a direct and alternating source of current is depicted in Fig. 1.1D. Heating of the wire produces an expansion of the air in the vicinity of the wire. Therefore, if the wire is heated and allowed to cool at an alternating rate, a sound wave will be produced. The heated conductor has been used in the calibration of microphones. Another example of a thermal sound generator is the arc shown in Fig. 1.1E. The arc has been used as the driver in a horn loudspeaker. Lightning is an example of an explosive-type arc sound generator. The result is thunder with an original wave form of the type shown in Fig. 1.1C.

Sound is produced when wind blows over wires as shown in Fig. 1.1F. These sounds are referred to as vortex or Aeolian-type tones. The generation of the sounds is produced by the nonlinearity in the air blowing past small or sharp obstacles.

1.3 NATURE AND EQUATIONS OF SOUND WAVES

A. Velocity of Propagation of a Sound Wave

The sound waves generated by the various means described in the preceding section are propagated with a definite velocity. The velocity of propagation is given by:

$$c = \sqrt{\frac{p_0 \gamma}{\rho}} \qquad (1.1)$$

where c = velocity of sound, in centimeters per second,

   γ = ratio of specific heats (1.4 for air),

   p_0 = static pressure, in microbars or dynes per square centimeter, and

   ρ = density, in grams per cubic centimeter.

The velocity of sound in air at 20° C at sea level is 344 meters per second. The density of air is 0.00129 grams per cubic centimeter.

B. Frequency of a Sound Wave

The continuous sound generators described in the preceding section produce recurrent waves. A complete set of these recurrent waves constitute a cycle. The number of recurrent waves or cycles which pass a certain observation point per second is termed the frequency of the sound wave. The unit of frequency is the hertz.

C. Wavelength of a Sound Wave

The wavelength of a sound wave is defined as the distance the sound travels to complete one cycle. The frequency of a sound wave is the number of cycles which pass a certain observation point per second. Therefore, the wavelength is the ratio of the velocity of propagation divided by the frequency, which is expressed as:

$$\lambda = \frac{c}{f} \qquad (1.2)$$

where λ = wavelength, in centimeters,

   c = velocity of sound, in centimeters per second, and

   f = frequency, in hertz.

D. Pressure in a Sound Wave

A sound wave consists of pressures above and below the normal undisturbed pressure in the air.

   The instantaneous sound pressure at a point is the total instantaneous pressure at that point minus the static pressure, the static pressure being the normal atmospheric pressure in the absence of sound. The unit is the dyne per square centimeter, or the microbar.

   The effective sound pressure at a point is the root-mean-square value of the instantaneous sound pressure over a complete cycle at that point. The unit is the dyne per square centimeter or the microbar. The term "effective sound pressure" is frequently shortened to "sound pressure."

The sound pressure in a spherical sound wave falls off inversely as the distance from the sound source increases.

E. Particle Velocity in a Plane Sound Wave

The passage of a sound wave produces a displacement of the particles or molecules in the gas from the normal position, that is, the position in the absence of a sound wave. The particle displacement in a normal sound wave in speech and music is a very small fraction of a millimeter. For example, in normal conversational speech at a distance of 10 feet from the speaker, the particle amplitude is of the order of one-millionth of an inch. The particle or molecule in the medium oscillates at the frequency of the sound wave. The velocity of a particle or molecule in the process of being displaced at the frequency of the sound wave is termed the particle velocity. The unit is the centimeter per second.

The relation between sound pressure and particle velocity is given by:

$$u = \frac{p}{\rho c} \tag{1.3}$$

where u = particle velocity, in centimeters per second,

p = sound pressure, in dynes per square centimeter or microbars,

ρ = density of air, in grams per square centimeter, and

c = velocity of sound, in centimeters per second.

F. Amplitude in a Plane Sound Wave

The amplitude or displacement of the particle from its position in the absence of a sound wave is given by:

$$d = \frac{u}{2\pi f} \tag{1.4}$$

where d = particle amplitude, in centimeters,

u = particle velocity, in centimeters per second, and

f = frequency, in hertz.

G. Plane Sound Wave

The equation for the pressure in a simple harmonic plane sound wave traveling in the positive x direction is given by:

$$p = akc^2\rho \sin k(ct - x) \tag{1.5}$$

where p = pressure, in microbars,

      a = amplitude of the particle displacement, in centimeters,

      $k = 2\pi/\lambda$,

      λ = wavelength, in centimeters,

      c = velocity of sound, in centimeters per second,

      ρ = density of air, in grams per cubic centimeter,

      t = time, in seconds, and

      x = distance of the observation point along the x axis from $x = 0$, in centimeters.

The equation for the particle velocity u, in centimeters per second, in a simple plane sound wave traveling in the positive x direction is given by:

$$u = akc \sin k(ct - x) \tag{1.6}$$

Comparing equations (1.5) and (1.6), one can see that the pressure and particle velocity in a plane sound wave are in phase.

The ratio of the pressure to the particle velocity in a plane sound wave is:

$$\frac{p}{u} = \rho c \tag{1.7}$$

The quantity ρc is the characteristic acoustical resistance of the medium.

H. Spherical Sound Wave

The equation for the pressure p, in dynes per square centimeter, in a spherical sound wave generated by a small sound source is given by:

$$p = \frac{S\rho ck}{4\pi r} \sin k(ct - r) \tag{1.8}$$

where p = pressure, in microbars,

      S = strength of the source, that is, the maximum rate of fluid emission of the small source, in cubic centimeters per second,

      r = distance from the origin, in centimeters; and all the other quantities as defined in Equation (1.5).

Equation (1.8) shows that the pressure varies inversely as the distance from the source.

The equation for the particle velocity u, in centimeters per second, in a spherical sound wave generated by a small sound source is given by:

$$u = -\frac{Sk}{4\pi r}\left[\frac{1}{kr}\cos k(ct-r) - \sin k(ct-r)\right] \qquad (1.9)$$

Equation (1.9) shows that the particle velocity in a spherical sound wave is an inverse function of r and r^2. The pressure and particle velocity are not in phase in a spherical sound wave save for large distances from the sound source when the spherical wave becomes a plane wave.

I. Phase Angle Between the Pressure and Particle Velocity in a Spherical Sound Wave

Comparing Equation (1.9) for the particle velocity and Equation (1.8) for the pressure in a spherical sound wave, the phase angle between the pressure and the particle velocity in a spherical sound wave is given by:

$$\theta = \tan^{-1}\frac{1}{kr} \qquad (1.10)$$

For very large values of kr, that is, a plane wave, the pressure and particle velocity are in phase. The phase angle θ as a function of kr is depicted in Fig. 1.2.

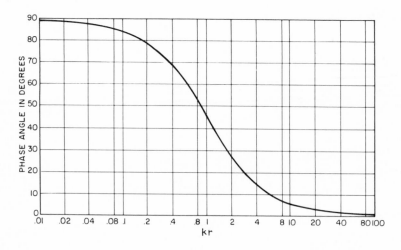

Fig. 1.2 Phase angle between the pressure and particle velocity in a spherical sound wave in terms of kr, where $k = 2\pi/\lambda$, λ = wavelength and r = distance from the sound source.

J. Ratio of the Absolute Magnitude of the Particle Velocity and the Pressure in a Spherical Sound Wave

Employing Equation (1.9) for the particle velocity and Equation (1.8) for the pressure in a spherical sound wave, the ratio of the absolute magnitude of the particle velocity to the absolute value of the pressure in a spherical sound wave is given by:

$$\text{Ratio} = \frac{\sqrt{1 + k^2 r^2}}{\rho c k r} \tag{1.11}$$

The normalized ratio of the particle velocity to the pressure in a spherical sound wave is depicted in Fig. 1.3.

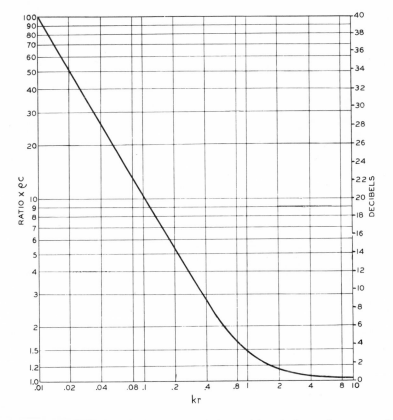

Fig. 1.3   Ratio of the absolute magnitude of the particle velocity to the pressure in a spherical sound wave in terms of kr, where $k = 2\pi/\lambda$, λ = wavelength and r = distance from the sound source.

K. Intensity or Power in a Sound Wave

The sound intensity of a sound field in a specified direction at a point is the sound energy transmitted per unit of time in the specified direction through a unit area normal to this direction at the point. The unit is the erg per second per square centimeter. It may also be expressed in watts per square centimeter.
The intensity of a plane wave is given by:

$$I = \frac{p^2}{\rho c} = pu = \rho c u^2 \tag{1.12}$$

where I = intensity, in ergs per second per square centimeter,

p = pressure, in dynes per square centimeter or microbars,

u = particle velocity, in centimeters per second,

c = velocity of propagation, in centimeters per second, and

ρ = density of the medium, in grams per cubic centimeter.

The product ρc is termed the specific acoustical resistance of the medium.

L. Sound Power Emitted by a Simple Point Source

A point source is a source small compared to the wavelength which alternately injects air into the air medium and withdraws air from the air medium. The sound power emitted by a simple source is given by:

$$P = \frac{\rho c k^2 U^2}{8\pi} \tag{1.13}$$

where P = sound power, in ergs per second,

U = maximum volume current, that is, the maximum rate of fluid emission, in cubic centimeters per second (for a sine wave the maximum amplitude is 1.4 times the root mean square amplitude),

ρ = density, in grams per cubic centimeter,

c = velocity of sound, in centimeters per second,

$k = 2\pi/\lambda$, and

λ = wavelength, in centimeters.

The sound power emitted by a simple source can also be expressed as a function of the volume displacement as follows:

$$P = \frac{\pi \rho c k^2 f^2 X^2}{2} = \frac{2\pi^3 \rho f^4 X^2}{c} \tag{1.14}$$

where P = sound power, in ergs per second,

X = maximum volume displacement, in cubic centimeters,

ρ = density, in grams per cubic centimeters,

$k = 2\pi/\lambda$,

c = velocity of sound, in centimeters per second,

λ = wavelength, in centimeters, and

f = frequency, in hertz.

The sound power output from a simple source as a function of the frequency for various displacements is shown in Fig. 1.4. The data of Fig. 1.4 illustrate the difficulties in obtaining sound power output in the low frequency range. For example, to develop an acoustic watt of output at 30 hertz requires the emission and alternate withdrawal of 2500 cubic centimeters of air per cycle. On the other hand, to develop an acoustic watt of output at 15,000 hertz requires the displacement of only 0.01 cubic centimeter of air per cycle.

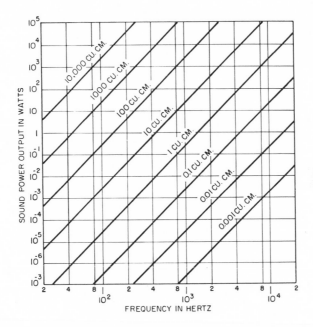

Fig. 1.4 Sound power output from a simple source in terms of the maximum volume displacement.

M. Point Source Radiating into a Semi-Infinite Medium. Solid Angle of 2π Steradians

The above examples considered a point source operating in an infinite medium as shown in Fig. 1.5A. The next problem of interest is that of a point source operating in a semi-infinite medium, for example, a point source near an infinite wall as depicted in Fig. 1.5B.

In this case we can employ the principle of images as shown in Fig. 1.5B. The pressure, assuming the same distance from the source, is two times that of the infinite medium. The particle velocity is also two times that of the infinite medium. The average power transmitted through a unit area is four times that of the infinite medium. The average power output of the source, however, is two times that of a simple source operating in an infinite medium.

N. Point Source Radiating into a Solid Angle of π Steradians

Employing the method of images in Fig. 1.5C the pressure is four times, the particle velocity is four times, and the average power transmitted through a unit area is sixteen times that of an infinite medium for the same distance. The average power output of the source is four times that of a simple source operating in an infinite medium.

O. Point Source Radiating into a Solid Angle of $\pi/2$ Steradians

Employing the method of images in Fig. 1.5D the pressure is eight times, the particle velocity eight times, and the average power transmitted through a unit area is sixty-four times that of the same source operating in an infinite medium at the same distance. The average power output is eight times that of the same simple source operating in an infinite medium.

P. Application of the Simple Source

The above data may be applied to acoustic radiators in which the dimensions are small compared to the wavelength and located close to the boundaries indicated above. For example, Fig. 1.5A would correspond to a loudspeaker, which acts as a simple source, suspended in space at a large distance from any walls or boundaries; Fig. 1.5B would correspond to a loudspeaker placed on the floor in the center of the room; Fig. 1.5C would correspond to a loudspeaker placed on the floor along a wall; and Fig. 1.5D would correspond to a loudspeaker placed in the corner of the room. Of course, as

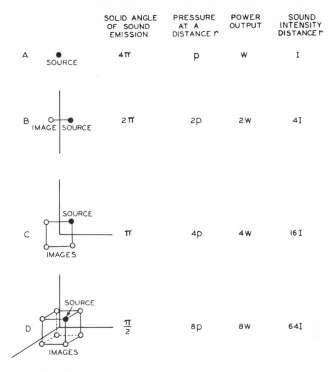

		SOLID ANGLE OF SOUND EMISSION	PRESSURE AT A DISTANCE r	POWER OUTPUT	SOUND INTENSITY DISTANCE r
A	SOURCE	4π	p	W	I
B	IMAGE SOURCE	2π	2p	2W	4I
C	SOURCE IMAGES	π	4p	4W	16I
D	SOURCE IMAGES	$\dfrac{\pi}{2}$	8p	8W	64I

Fig. 1.5 The sound pressure, total power output and energy density delivered by a point source operating in solid angles of 4π, 2π, π and $\pi/2$ steradians.

pointed out above, these examples hold only when the dimensions of the radiator and the distance from the wall are small compared to the wavelength.

Q. Acoustic Impedance

Acoustic impedance is the complex ratio of the sound pressure and volume velocity. In the case of the acoustic impedance per unit area, termed the specific acoustic impedance, the volume current becomes the particle velocity. Employing Equation (1.5) for the pressure and Equation (1.6) for the particle velocity, the specific acoustic impedance for a plane wave is the following:

$$z_1 = \frac{p}{u} = \rho c \qquad (1.15)$$

where z_1 = specific acoustic impedance (resistance) in acoustic ohms.

In the case of a complex acoustic impedance the expression is given by:

$$z_A = r_A + x_A \qquad (1.16)$$

where

z_A = acoustic impedance, in acoustic ohms,

r_A = acoustic resistance, in acoustic ohms, and

x_A = acoustic reactance, in acoustic ohms.

R. Acoustic Impedance Presented by the Air Load Upon Vibrating Pistons

Practically all of the radiators or vibrators in sound-reproducing systems are circular pistons. Therefore, it seems appropriate to provide the acoustic impedance of the air load upon a vibrating piston under three conditions of operation as depicted in Fig. 1.6.

The acoustical impedance may be obtained from the data of Fig. 1.6 as follows:

$$z_A = \frac{\rho c(r_1 + x_1)}{S} \qquad (1.17)$$

where

z_A = acoustical impedance presented to one side of the piston, in acoustic ohms,

r_1 = acoustic resistance per unit area divided by ρc, as depicted in the graph of Fig. 1.6,

x_1 = acoustic reactance per unit area divided by ρc, as depicted in the graph of Fig. 1.6, and

S = area of the piston, in square centimeters.

S. Decibels

In acoustics the ranges of intensities, pressures, etc., are so large that it is convenient to use a scale of smaller numbers termed decibels. The abbreviation dB is used for the term decibel. The bel is the fundamental division of a logarithmic scale for expressing the ratio of two amounts of power, the number of bels denoting such a ratio being the logarithm to the base ten of this ratio. The decibel is one tenth of a bel. For example, with P_1 and P_2 designating two amounts of power and n the number of decibels denoting their ratio:

$$n = 10 \log_{10} \frac{P_1}{P_2} \text{ decibels} \qquad (1.18)$$

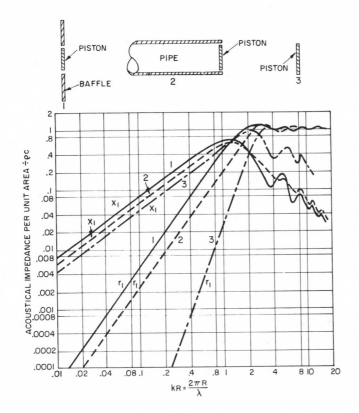

Fig. 1.6 Acoustic impedance of the air load upon pistons operating under various conditions. 1. Piston in an infinite baffle. 2. Piston in the end of an infinite pipe. 3. Piston in free space.

When the conditions are such that ratios of currents or ratios of voltages (or the analogous quantities such as pressures, volume currents, forces, and particles velocities) are the square roots of the corresponding power ratios, the number of decibels by which the corresponding powers differ is expressed by the following formulas:

$$n = 20 \log_{10} \frac{i_1}{i_2} \text{ decibels} \tag{1.19}$$

$$n = 20 \log_{10} \frac{e_1}{e_2} \text{ decibels} \tag{1.20}$$

where i_1/i_2 and e_1/e_2 are the given current and voltage ratios, respectively.

For the relation between decibels and power and current or voltage ratios, see Table 1.1.

TABLE 1.1

The Relation Between Decibels and Power and Current or Voltage Ratios

Power Ratio	Decibels	Current or Voltage Ratio	Decibels
1	0	1	0
2	3.0	2	6.0
3	4.8	3	9.5
4	6.0	4	12.0
5	7.0	5	14.0
6	7.8	6	15.6
7	8.5	7	16.9
8	9.0	8	18.1
9	9.5	9	19.1
10	10	10	20
100	20	100	40
1,000	30	1,000	60
10,000	40	10,000	80
100,000	50	100,000	100
1,000,000	60	1,000,000	120

T. Diffraction of Sound

Diffraction is the change in direction of propagation of sound due to the passage of sound around an obstacle. The passage of sound around an obstacle is a well-known occurrence. The larger the ratio of the wavelength to the dimensions of the obstacle the greater the diffraction. The diffraction around the human head is important in both speaking and listening. The diffraction of sound by microphones and loudspeakers is important in the performance of those instruments. Figure 1.7 shows the diffraction of sound by a sphere, a cube, and a cylinder as a function of the ratio of dimensions to wavelength. The data of Fig. 1.7 may be used to predict the diffraction by objects of these general shapes. For example, the sphere may be used to predict the diffraction of sound by the human head. The cylinder may be used to predict the diffraction of sound by microphones of a cylindrical shape. The cube may be used to predict the diffraction of sound by a loudspeaker in a cubical cabinet.

1.4 SYSTEM OF UNITS

The International Standard of Units ISO Recommendation R1000 is based on the meter, kilogram, and second. However, in development and research

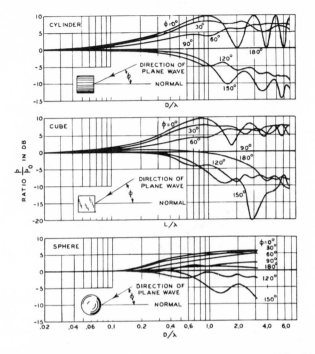

Fig. 1.7   The diffraction of sound by a cylinder, cube and sphere. (After Muller, Black and Davis.)

on sound-reproducing systems, the centimeter, gram, and second are more convenient and are still used by a large number of investigators. For example, sound pressure is almost universally expressed in microbars, that is, dynes per square centimeter. For the acoustics of rooms, studios, theaters, etc., almost all of the investigations employ the foot, square foot, cubic foot, etc.

Accordingly, the system of units employed in this book are those in most common use. If the reader wishes to use other units, conversion is a simple matter.

REFERENCES

Muller, G. G., Black, R., and Davis, T. E. "The Diffraction Produced by Cylindrical and Cubical Obstacles and by Circular and Square Plates," *Journal of the Acoustical Society of America*, Vol. 10, No. 1, p. 6, 1938.

Olson, Harry F. *Acoustical Engineering*, Van Nostrand Reinhold, New York, 1957.

Olson, Harry F. "Direct Radiator Loudspeaker Enclosures, Reissue," *Journal of the Audio Engineering Society*, Vol. 17, No. 1, p. 22, 1969.

2
Loudspeakers

2.1 INTRODUCTION

A loudspeaker is an electroacoustic transducer designed to radiate acoustic power into the air with the resultant acoustic waveform being essentially equivalent to that of the electrical input. There are two general types of loudspeakers in use today, namely, the direct radiator and the horn loudspeakers. The diaphragm of the direct radiator loudspeaker is coupled directly to the air. The diaphragm of the horn loudspeaker is coupled to the air by means of a horn. The direct radiator and horn loudspeakers are used separately and in combinations of direct radiator and horn types. The purpose of this chapter is to describe the constructional features and performance characteristics of direct radiator and horn-type loudspeakers and combinations of the two types.

2.2 DIRECT RADIATOR DYNAMIC LOUDSPEAKER

The almost universal use of the direct radiator dynamic loudspeaker is due to simplicity of construction, small space requirements, and a relatively uniform frequency response characteristic. Uniform response over a moderate frequency band may be obtained with any simple direct radiator dynamic loudspeaker. The reproduction over a wide frequency range by a single unit is limited by practical considerations. However, the use of two or more direct radiator dynamic mechanisms and appropriate enclosure design will provide uniform response over the entire audio frequency range.

A. Direct Radiator Dynamic Loudspeaker Mechanism

A simple dynamic direct radiator loudspeaker mechanism consists of a cone-type diaphragm coupled to a voice coil located in the air gap of a magnetic structure as depicted in Fig. 2.1. The essential elements of the direct radiator dynamic loudspeaker mechanism are the cone, voice coil, suspensions, and magnetic field structure. These elements will be described in the sections which follow.

B. Cones and Suspensions

The diaphragm or cone of practically all direct radiator dynamic loud-speakers is made of paper. The cones shown in Fig. 2.2 are made by a felting process employing a master screen having the shape of the diaphragm. A mixture of pulp and water is drawn through the screen leaving a thin deposit of compressed pulp. The deposit is dried and removed from the screen. The result is the finished diaphragm. The outside suspension can also be felted as an integral part of the cone.

There are two types of felted cones in general use, namely, the circular type of Fig. 2.2A and the elliptical type of Fig. 2.2B. In certain cabinets or

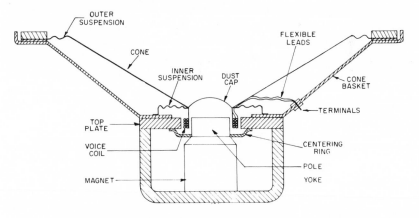

Fig. 2.1   Sectional view of a direct radiator dynamic loudspeaker.

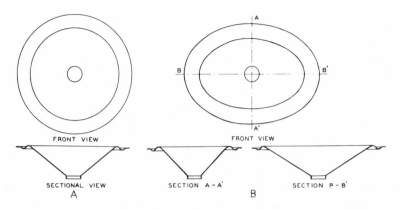

Fig. 2.2 Felted paper cones for direct radiator dynamic loudspeakers. A. Circular cone. B. Elliptical cone.

enclosures it is possible to obtain a larger cone area by employing the elliptical cone. The directional pattern of the elliptical cone is sharper in the plane containing the major axis and the axis of the cone, and is broader in the plane containing the minor axis and the axis of the cone than for the circular cone of the same area.

Three types of cross sections of cones are depicted in Fig. 2.3. The shapes are the conical of Fig. 2.3A, the flared of Fig. 2.3B, and the corrugated conical of Fig. 2.3C. The shapes of Fig. 2.3 can be employed in either the circular or elliptical types. The flared shape is more rigid than the conical shape; for this reason the directional pattern in the high frequency range is somewhat sharper than for the conical type. The use of corrugations increases the radial rigidity and slows the propagation of the sound wave in the cone and thereby broadens the directional pattern.

A second cone, termed the whizzer, may be employed as shown in Fig. 2.4A to extend the high frequency range. The compliance in the main cone uncouples the main cone in the high frequency range, and all the driving force is applied to the whizzer cone in the high frequency range.

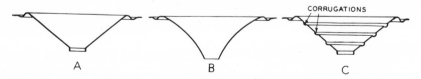

Fig. 2.3 Sectional views of felted paper cones for direct radiator dynamic loudspeakers. A. Conical shape. B. Flared shape. C. Conical shape with corrugations.

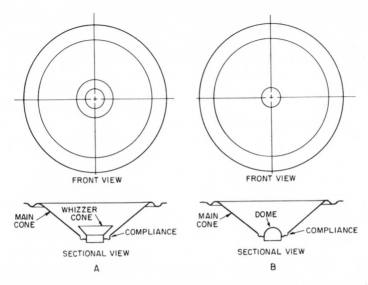

Fig. 2.4 Sectional views of felted paper cones with adjunct high frequency radiators. A. A small cone termed a whizzer. B. A domed dust cap.

The domed dust cap depicted in Fig. 2.4B may also be used to extend the high frequency range by including the corrugation in the main cone.

Three types of outside cone suspension are depicted in Fig. 2.5. The most common suspension is the one felted integral with the cone as shown in Fig. 2.5A. If the excursions of the cone are very large, the felted paper suspension may deteriorate and ultimately fail. This may be overcome by the use of the phenolic impregnated cloth suspension of Fig. 2.5B. This suspension is cemented to the cone as shown in Fig. 2.5B. To provide a very flexible suspension a flat pliable plastic is used as depicted in Fig. 2.5C.

There are many designs of inner or centering suspensions. The most common type consists of a multi-corrugated disk of phenolic impregnated cloth as shown in Fig. 2.1.

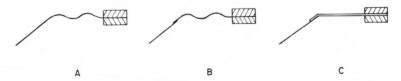

Fig. 2.5 Sectional views of cone suspension systems. A. Corrugated suspension felted integral with the cone. B. Corrugated phenolic impregnated cloth cemented to the cone. C. Flexible flat plastic cemented to the cone.

C. Magnetic Field Structures

Five typical magnetic field structures are shown in Fig. 2.6. A magnetic field structure with a cylindrical magnet and an iron center pole is shown in Fig. 2.6A. The air gap flux densities which can be obtained with this structure are limited by the flux density capacities of the iron magnetic structure since there are no practical limitations on the dimensions of the magnet. For soft iron the flux density in the air gap is limited to 15,000 gausses. For an iron–cobalt alloy the flux density in the air gap is limited to 22,000 gausses.

A center pole magnet in a yoke-type iron magnetic field return structure is shown in Fig. 2.6B. A center pole magnet in a cup type iron magnetic field return structure is shown in Fig. 2.6C. The maximum flux density in the air gap of the magnetic field structures of Figs. 2.6B and 2.6C is 10,000 gausses, as determined by flux-carrying capacity of the magnet.

The iron structures of Figs. 2.6D and 2.6E are similar to the magnetic field structures of Figs. 2.6B and 2.6C. However, the cross section of the center magnet is larger in order to obtain a high flux density in the air gap. In order to accommodate the larger magnet, a center pole of iron is used. In this structure the limitation on the flux density in the air gap is determined by the flux-carrying capacity of the iron. Employing soft iron, flux densities up to 15,000 gausses can be obtained in the air gap. Using an iron–cobalt alloy, flux densities up to 22,000 gausses can be obtained in the air gap. However, the very high flux densities are seldom used in direct radiator dynamic loudspeakers. (See Section 2.2E.)

D. Voice Coils

Voice coil designs used in direct radiator dynamic loudspeakers are shown in Fig. 2.7. In Fig. 2.7A the axial length of the voice coil is shorter than the

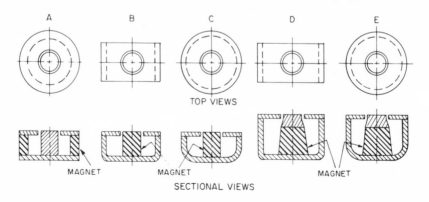

Fig. 2.6 Top and sectional views of permanent magnet field structures.

Fig. 2.7 Sectional views of voice coil designs for dynamic direct radiator loudspeakers

axial length of the air gap. In this structure the voice coil remains in a high flux density as long as the excursion of the coil does not exceed the limits of the air gap. In Fig. 2.7B the axial length of the coil is the same as the axial length of the air gap. If the excursion of the coil extends beyond the air gap, the flux-density–voice-coil-turns product is not a constant. Under these conditions, the driving force does not correspond to the current in the coil, and the result is the production of nonlinear distortion. This design of voice coil can be used where the amplitudes are very small, as for example, in the high frequency range. In Fig. 2.7C the axial length of the voice coil is greater than the axial length of the air gap. In this design the flux-density–flux-turns product remains a constant for even very large amplitudes. There will be no nonlinear distortion generated in the voice coil magnetic field interaction if the voice coil does not exceed the air gap limits. However, the efficiency of the voice coil of Fig. 2.7C is low because a large part of the voice coil is always out of the magnetic field of the air gap. The design of Fig. 2.7C is used in low frequency loudspeakers where the amplitude of the cone is large.

E. Complementary Voice Coil–Suspension Design

The compliance of the suspension system of a dynamic direct radiator loudspeaker is nonlinear. The compliance of the suspension system decreases with increase in amplitude as shown in Fig. 2.8. The voice coil can be designed so that the nonlinear effects of the driving force developed in the voice coil counters the nonlinear effects of the suspension.

The compliance of a typical loudspeaker suspension system as a function of the displacement of the cone is shown in Fig. 2.8. The compliance may be approximately expressed as:

$$C_{Mx} = \frac{C_{M0}}{1 + \beta x^2} \tag{2.1}$$

where C_{Mx} = compliance for a displacement x, in centimeters per dyne,
C_{M0} = compliance for $x = 0$,
x = displacement of the cone in centimeters, and
β = constant.

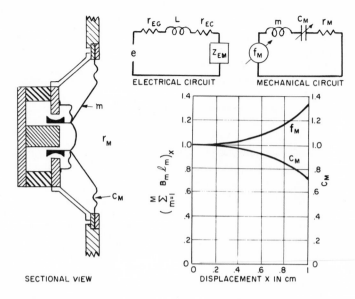

ELECTRICAL CIRCUIT   MECHANICAL CIRCUIT

SECTIONAL VIEW   DISPLACEMENT X IN cm

Fig. 2.8 Sectional view, electrical circuit and mechanical circuit of the vibrating system of a direct radiator dynamic loudspeaker mechanism with a nonlinear suspension and a complementary nonlinear driving system. In the electrical circuit: e = internal voltage of the generator. r_{EG} = the internal electrical resistance of the generator. r_{EC} and L = electrical resistance and inductance of the voice coil. z_{EM} = the motional electrical impedance. In the mechanial circuit: m = mass of the cone, voice coil and air load. C_M = compliance of the suspension system. r_M = mechanical resistance of the air load and suspension system. f_M = driving force. The graph depicts the flux density-voice coil turns-product, labelled f_M and the compliance, labelled C_M.

A voice coil design with a flux-density–voice-coil-turns product which complements the compliance for large amplitudes is shown in Fig. 2.8. The concentration of turns increases with the axial distance from the axial center of the voice coil as shown in Fig. 2.8. Specifically to counter the nonlinearity of the compliance, the flux-density–voice-coil-turns product is designed to follow the expression:

$$\sum_{m=1}^{M} (B_m l_m)_x = \sum_{m=1}^{M} (B_m l_m)_0 (1 + \gamma x^2) \qquad (2.2)$$

where B_m = flux density associated with turn l_m, in gausses,
            l_m = length of the conductor in turn m, in centimeters,
            M = total number of turns in the voice coil, and
            γ = constant.

The summation on the right-hand side of Equation 2.2 is for $x = 0$.

The flux-density–voice-coil-turns product for the voice coil of Fig. 2.8, as a function of the displacement, is shown in Fig. 2.8.

The differential equation for the system of Fig. 2.8 becomes:

$$m\ddot{x} + r_M\dot{x} + \frac{1 + \beta_1 x^2}{C_{MO}} x = \left(\sum_{m=1}^{M} B_m l_m \right)_0 (1 + \gamma_1 x^2)i \cos \omega t \qquad (2.3)$$

where m = mass of the cone, voice coil, and air load in grams,
r_M = mechanical resistance of the air load and suspension, in mechanical ohms,
i = amplitude of the current in the voice coil, in abamperes,
$\omega = 2\pi f$, and
f = frequency, in hertz.

A consideration of the nonlinear differential equation (2.3) shows that the two nonlinear distortions cancel each other. A reduction of the nonlinear distortion of about one-fourth that of the extended voice coil of Fig. 2.7C can be achieved by the complementary system. However, there is an even further reduction in efficiency beyond that of the extended voice coil design.

F. Performance Considerations

A sectional view, voice coil electrical circuit, and mechanical circuit of a direct radiator dynamic loudspeaker is shown in Fig. 2.9. The determination of the performance characteristics that follows involves the system depicted in Fig. 2.9.

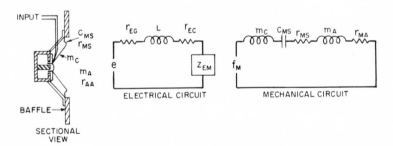

Fig. 2.9 Sectional view, electrical circuit and mechanical circuit of the vibrating system of a direct radiator dynamic loudspeaker mechanism mounted in a very large baffle. In the electrical circuit, e = internal voltage of the generator. r_{EG} = internal electrical resistance of the generator. r_{EC} and L = electrical resistance and inductance of the voice coil. z_{EM} = motional electrical impedance. In the mechanical circuit, m_C = mass of the cone and voice coil. C_{MS} = compliance of the suspension system. r_{MS} = mechanical resistance of the suspension system. m_A = mass of the air load. r_{MA} = mechanical resistance of the air load. f_M = driving force.

The interaction of the current in the voice coil with the magnetic field produces a corresponding force which is transmitted to the cone. The force generated in the voice coil is given by:

$$f_M = Bli \tag{2.4}$$

where f_M = force, in dynes,
B = flux density in the air gap, in gausses,
l = length of the voice coil conductor, in centimeters, and
i = current in the voice coil, in abamperes.

The current in the voice coil can be determined from the electrical circuit of Fig. 2.9 as follows:

$$i = \frac{e}{r_{EG} + r_{EC} + jwL + z_{EM}} \tag{2.5}$$

where e = voltage of the generator, in abvolts,
r_{EG} = electrical resistance of the generator, in abohms,
r_{EC} = electrical resistance of the voice coil, in abohms,
L = inductance of the voice coil, in abohms, and
z_{EM} = electrical motional impedance due to the mechanical system, in abohms.

The electrical motional impedance is given by:

$$z_{EM} = \frac{(Bl)^2}{z_M} \tag{2.6}$$

where z_{EM} = electrical motional impedance, in abohms,
B = flux density in the air gap, in gausses,
l = length of the voice coil conductor, in centimeters, and
z_M = mechanical impedance at f_M in Fig. 2.9, in mechanical ohms.

In the frequency range above the resonant frequency the mechanical elements that involve the velocity of the cone are the mass of the cone and voice coil and the mass due to the air load. Under these conditions the velocity of the cone and coil is given by:

$$\dot{x} = \frac{f_M}{j\omega(m_C + m_A)} \tag{2.7}$$

where $\dot{x}$ = velocity of the coil and cone, in centimeters per second,
f_M = force given by Equation (2.4),
m_C = mass of the cone and voice coil, in grams,
m_A = mass of the air load, in grams, as given by Equation (2.3),
$\omega = 2\pi f$, and
f = frequency, in hertz.

The mass of the air load on one side of the cone is given by:

$$m_A = \frac{x_1 S \rho c}{2\pi f} \qquad (2.8)$$

where m_A = mass of the air load, in grams,
x_1 = acoustic reactance of the air load per unit area divided by ρc as depicted in the graph in Fig. 1.6,
S = area of the cone, in square centimeters,
f = frequency, in hertz,
ρ = density of air, in grams per cubic centimeter, and
c = velocity of sound, in centimeters per second.

The sound power output of the cone is given by:

$$P = \rho c r_1 S \dot{x}^2 \qquad (2.9)$$

where P = sound power output, in ergs per second,
r_1 = acoustic radiation resistance per unit area divided by ρc as depicted in the graph in Fig. 1.6,
ρ = density of air, in grams per cubic centimeter,
c = velocity of sound, in centimeters per second,
S = area of the cone, in square centimeters, and
$\dot{x}$ = velocity of the cone, in centimeters per second, as given by by Equation (2.7).

Referring to Equation (2.4), the force is constant for constant current in the voice coil. Then, from Equation (2.7), the velocity is inversely proportional to the frequency. However, the radiation resistance of Fig. 1.6 is proportional to the square of the frequency in the frequency region below the ultimate radiation resistance. Therefore, it follows that from Equation (2.9), the second power output is independent of the frequency in this frequency region. These facts constitute the essential features of the direct radiator dynamic loudspeaker which conspire to produce a uniform response frequency characteristic.

The efficiency of a loudspeaker is the ratio of the sound power output to the electrical power input. The efficiency of a direct radiator dynamic loudspeaker mounted in an infinite baffle or wall is given by:

$$\mu = \frac{B^2 r_1 \rho c S m}{[(\omega m_C + \omega m_A)^2 \rho_c K_r] 10^3} \times 100 \qquad (2.10)$$

where μ = efficiency, in percent,
B = flux density in the air gap, in gausses,
r_1 = acoustic radiation resistance per unit area divided by ρc, as depicted in the graph in Fig. 1.6,

ρ = density of air, in grams per cubic centimeter,

S = area of the cone, in square centimeters,

m = mass of the voice coil, in grams,

m_C = mass of the cone and coil, in grams,

ω = $2\pi f$,

f = frequency, in hertz,

c = velocity of sound, in centimeters per second,

m_A = mass of the air load, in grams, as given by Equation (2.8),

ρ_c = density of the conductor of the voice coil, in grams per cubic centimeter, and

K_r = resistivity of the voice coil conductor, in microhms per centimeter cube.

Equation (2.10) shows that the efficiency is proportional to the square of the flux density. The flux density in the air gap of commercial direct radiator loudspeakers varies from 3000 to 15,000 gausses, a situation which leads to a wide variation in efficiency.

Copper and aluminum wire are used in the construction of the voice coils. For copper the product $\rho_c K_r$ is 15.2 in grams per cubic centimeter times microhms per centimeter cube and for aluminum $\rho_c K_r$ is 7.6 in grams per cubic centimeter times microhms per centimeter cube. Therefore, if the flux density in the air gap remains a constant, Equation (2.10) shows that the efficiency of a loudspeaker with an aluminum voice coil will be twice that of a loudspeaker with a copper voice coil. However, the volume of the air gap must be larger to accommodate the aluminum voice coil. There is greater difficulty in obtaining the same efficiency in the high frequency range compared to the low frequency range. Therefore, aluminum voice coil loudspeakers are used for the high frequency ranges and copper voice coil loudspeakers are used for the low frequency ranges.

The efficiency of a high-quality commercial direct radiator dynamic loudspeaker is of the order of 3 to 5 percent. For some of the lower-grade direct radiator dynamic loudspeakers, the efficiency is a fraction of 1 percent.

G. Frequency Response Characteristics

Typical response frequency characteristics of direct radiator dynamic loudspeakers for cones with diameters of 16, 8, 4, 2, 1, and 0.5 inches are shown in Fig. 2.10. The data of Fig. 2.10 are for the operation of the loudspeaker mechanisms in an infinite baffle. The frequency response characteristics of Fig. 2.10 confirm the theoretical considerations of the preceding section. However, uniform response is maintained to some degree beyond the ultimate acoustic resistance of Fig. 1.6 because the effective mass of the cone is reduced

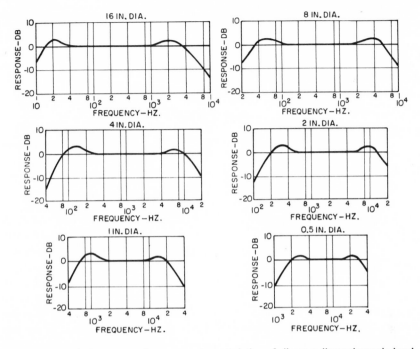

Fig. 2.10 Typical response frequency characteristics of direct radiator dynamic loudspeakers for diameters as shown.

in the high frequency region, the mass reactance of the air load decreases, and the directivity of the cone narrows with increase of the frequency.

H. Directional Characteristics

Typical directional characteristics of direct radiator dynamic loudspeakers are shown in Fig. 2.11. These characteristics show that the directivity sharpens as the ratio of the diameter to the wavelength increases. Certain deviations from the directivity patterns shown in Fig. 2.11 can be achieved by modifications of the cone design. However, the directivity patterns of Fig. 2.11 represent the directivity performance of typical cones of the type shown in Fig. 2.3A. The cone of Fig. 2.3B will exhibit a somewhat narrower directivity pattern than the cone of Fig. 2.3A for the larger ratios of the diameter to the wavelength. The cone of Fig. 2.3C will exhibit a somewhat broader directivity pattern than the cone of Fig. 2.3A for the larger ratios of diameter to the wavelength. The directivity patterns are broader than those of a piston of the same diameter. The cone does not behave as a piston due to the finite propagation of sound in the paper material of the cone.

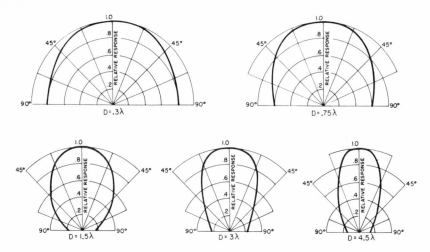

Fig. 2.11   Typical directional characteristics of dynamic direct radiator loudspeakers as a function of the ratio of the diameter of the cone to the wavelength for the cone shape of Fig. 2.3A.

I.   Sound Power Output

Employing Equation (2.9) the sound power output of a cone mounted in a large cabinet may be written in terms of the peak amplitude as follows:

$$P = (2\pi f)^2 \rho c r_1 S x^2 \tag{2.11}$$

where f = frequency, in hertz, and
   x = peak amplitude of the cone from the point of repose, in centimeters, and all the other quantities are the same as in Equation (2.9).

The peak amplitudes as a function of the frequency for cones with diameters of 16, 8, 4, 2, 1, and 0.5 inches for an acoustic output of 1 watt are shown in Fig. 2.12. The data of Fig. 2.12 shows that large sound power outputs cannot be obtained in the low frequency range from cones of small diameter because the amplitude becomes excessive.

J.   Baffles and Enclosures

Mounting and enclosure arrangements of direct radiator dynamic loudspeakers mechanisms are depicted in Fig. 2.13.

A direct radiator dynamic loudspeaker mechanism mounted in a flat baffle is shown in Fig. 2.13A. A relatively large baffle must be used to obtain adequate low frequency response because the baffle cutoff is determined by the

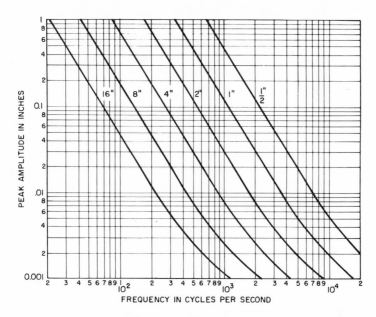

Fig. 2.12 The peak amplitude as function of the frequency for vibrating pistons (cones) of various diameters mounted in an infinite baffle, for one watt output on one side.

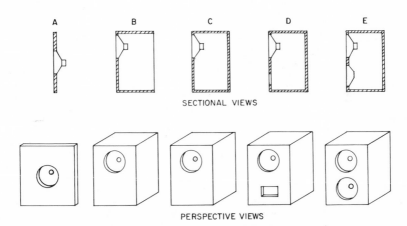

Fig. 2.13 Sectional and perspective views of mounting arrangements for direct radiator dynamic loudspeaker mechanisms. A. Flat baffle. B. Open back cabinet. C. Enclosed cabinet. D. Ported cabinet. E. Drone cone.

distance from the front to the back of the baffle. When this distance is less than a quarter of a wavelength, the response falls off rapidly below this frequency. To obtain good response down to 60 hertz requires a baffle $4\frac{1}{2} \times 4\frac{1}{2}$ feet. The low frequency response range cutoff is inversely proportional to the dimensions of the baffle. For example, a baffle $2\frac{1}{4} \times 2\frac{1}{4}$ feet will provide good response down to 120 cycles.

A direct radiator dynamic loudspeaker mechanism mounted in an openback cabinet is shown in Fig. 2.13B. The open-back cabinet is a combination baffle and resonator. The cabinet resonates as a pipe closed at one end and open at the other end. A cabinet with a height of 3 feet, a width of 2 feet and a depth of $1\frac{1}{2}$ feet will exhibit a resonant frequency of 80 hertz. The response is uniform down to where accentuation due to cabinet resonance occurs. The accentuation of response is of the order of 6 to 10 decibels. The response falls off quite rapidly below 60 hertz for a cabinet of these dimensions. As in the case of the baffle, low frequency response range cutoff and the resonance frequency of the cabinet are inversely proportional to the dimensions of the cabinet.

A direct radiator dynamic loudspeaker mechanism mounted in a completely enclosed cabinet is shown in Fig. 2.13C. The arrangement of Fig. 2.13C, employing a low resonant frequency mechanism, provides excellent response and low distortion when the compliance of the cabinet plays the major role in determining the resonant frequency as well as the controlling acoustic impedance in the low frequency range. Therefore, this system will be described in some detail.

A direct radiator dynamic loudspeaker mechanism with the back of the cone completely enclosed by the cabinet is shown in Fig. 2.14. At the low

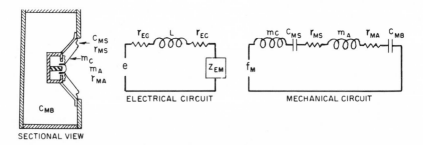

SECTIONAL VIEW   ELECTRICAL CIRCUIT   MECHANICAL CIRCUIT

Fig. 2.14 Sectional view, electrical circuit and mechanical circuit of the vibrating system of a direct radiator dynamic loudspeaker mechanism mounted in a closed back cabinet. In the electrical circuit, e = internal voltage of the generator. r_{EG} = internal electrical resistance of the generator. r_{EC} and L = electrical resistance and inductance of the voice coil. z_{EM} = motional electrical impedance. In the mechanical circuit, m_C = mass of the cone and voice coil. C_{MS} = compliance of the suspension system. r_{MS} = mechanical resistance of the suspension system. m_A = mass of the air load. r_{MA} = mechanical resistance of the air load. C_{MB} = compliance of the cabinet. f_M = driving force.

frequencies, the system is a simple source. (See Section 1.3L.) Under these conditions the mechanical radiation resistance is proportional to the square of the frequency up to the ultimate mechanical radiation resistance. The mechanical circuit of Fig. 2.14 shows that under these conditions, the output will be independent of the frequency above the resonant frequency of the system. (See Section 2.2F.)

A consideration of the mechanical circuit of Fig. 2.14 shows that the fundamental resonance is determined by the compliance of the suspension and the compliance of the enclosure. The compliance of the enclosure is given by:

$$C_{MB} = \frac{V}{\rho c^2 S_c^{\,2}} \tag{2.12}$$

where C_{MB} = compliance of the air in the cabinet as presented to the cone, in centimeters per dyne,

   V = volume of the cabinet, in cubic centimeters,

   S_c = area of the cone, in square centimeters,

   ρ = density of the air, in grams per cubic centimeter, and

   c = velocity of sound, in centimeters per second.

A consideration of Equation (2.12) and the mechanical circuit of Fig. 2.14 shows that the compliance of the cabinet can be made the controlling compliance for a low resonant frequency direct radiator dynamic loudspeaker mechanism.

Figure 2.15A gives the resonant frequency of 15-, 12-, and 8-inch direct radiator dynamic loudspeaker mechanisms as a function of the cabinet volume. When the resonant frequency has been derived from Fig. 2.15A, the response can be determined from Fig. 2.15B. When the compliance of the cabinet is the controlling acoustic impedance, the nonlinear distortion due to

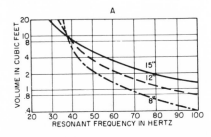
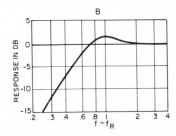

Fig. 2.15 A. The resonant frequency of 15-, 12- and 8-inch direct radiator loudspeaker mechanisms as a function of the cabinet volume. The free resonant frequency of the mechanisms are 15, 25 and 30 hertz for the 15-, 12- and 8-inch mechanisms, respectively. B. Response frequency characteristic of a direct radiator dynamic loudspeaker in terms of the resonant frequency f_R obtained from A.

suspension system of the mechanism is reduced to about one-fourth that of the same mechanism operating in a flat baffle. (See Section 2.5C.) The data of Fig. 2.15 show that excellent low frequency response may be obtained with a relatively small loudspeaker mechanism operating in a small cabinet. However, the power-handling capacity is limited by the permissible excursion of the cone. (See Section 2.21.)

A direct radiator dynamic loudspeaker mechanism mounted in a ported cabinet is shown in Fig. 2.13D. Compared to the completely enclosed cabinet, the ported cabinet will provide some accentuation in low frequency response but at the expense of a higher low frequency response range cutoff.

In Fig. 2.13E a drone cone is substituted for the port of Fig. 2.13D. The use of a drone cone provides additional acoustic radiation resistance with a corresponding improvement in low frequency performance as compared to the ported cabinet of Fig. 2.13D.

A loudspeaker which provides an omnidirectional directivity pattern in the horizontal plane is shown in Fig. 2.16A. The low frequency radiation from the cone is diffracted through the four apertures. The high frequency radiation from the cone is reflected through the four apertures by the cone-shaped reflector. The back of the cone is completely enclosed.

A column-type loudspeaker consisting of a line of small direct radiator dynamic loudspeaker mechanisms mounted in a cabinet is shown in Fig. 2.16B. The directivity pattern in the horizontal plane is the same as that shown in Fig. 2.11. However, the directivity pattern in the vertical plane is that of a series of sound sources in a line. Since the line is relatively long, the directivity pattern in the vertical plane is quite narrow. To cover the entire audio frequency range, two or three column loudspeakers are used. A uniform directivity pattern can be obtained in the vertical plane by employing a long

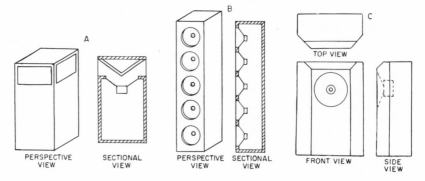

Fig. 2.16 A. Omnidirectional loudspeaker. B. Column loudspeaker. C. A cabinet with chamfered front corners.

column for the low frequency range, a medium-length column for the mid-frequency range and a short column for the high frequency range with suitable electrical cross-over networks. The column loudspeaker is useful in sound reenforcement where the radiated sound must be confined to a narrow beam in the vertical plane in order to increase the ratio of direct to reflected sound. (See Section 14.2F.)

Even though a direct radiator loudspeaker mechanism produces a uniform acoustical output with respect to frequency when mounted in a large flat baffle, the diffraction effects of the enclosure may introduce wide variations in the response. The sudden changes in acoustic impedance at the edges of the cabinet produces diffraction and reflection effects which interfere with the direct radiation from the loudspeaker with a resultant nonuniform frequency response. (See Section 1.2T.) These effects can be reduced to a negligible amount by employing a cabinet with large chamfered corners at the front corner edges of the cabinet, as shown in Fig. 2.16C.

K. Multiple Mechanisms, Multiple Channel

Multiple direct radiator dynamic loudspeaker mechanisms each operating in different frequency ranges are used to provide uniform response and directivity and adequate sound power output over the entire audio frequency range. The data of Figs. 2.10, 2.11, and 2.12 show that uniform response and directivity and adequate sound power output cannot be achieved by the use of a single loudspeaker mechanism.

A loudspeaker consisting of two direct radiator dynamic loudspeaker mechanisms and an electrical network for allocating the electrical input to the two mechanisms in the appropriate frequency ranges is shown in Fig. 2.17A.

A loudspeaker consisting of three direct radiator dynamic loudspeaker mechanisms and an electrical network for allocating the electrical input to the three mechanisms in the appropriate frequency ranges is shown in Fig. 2.17B.

In view of the separation between the mechanisms in Fig. 2.17, great care must be exercised in the cross-over frequency range in order to prevent interference between the mechanisms for observation points removed from the major axis of the system. These difficulties may be overcome in the axial-type systems shown in Fig. 2.18.

A loudspeaker consisting of two direct radiator dynamic loudspeaker mechanisms mounted axially and an electrical network for allocating the electrical input to the two mechanisms in the appropriate frequency ranges is shown in Fig. 2.18A.

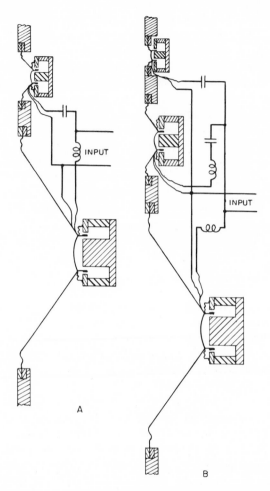

Fig. 2.17 Multiple direct radiator loudspeaker systems mounted in the front panel of the cabinet and the electrical networks. A. Two mechanism system. B. Three mechanism system.

A loudspeaker consisting of two direct radiator loudspeaker mechanisms mounted axially with the cones in the two mechanisms geometrically congruent and an electrical network for allocating the electrical input to the two mechanisms in the appropriate frequency ranges is shown in Fig. 2.18B. The overlap problems are reduced to a minimum in this design because the large cone is a geometrical continuation of the small cone so that in the frequency-overlap region the system operates as a single cone.

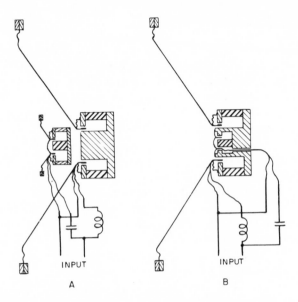

Fig. 2.18 Multiple direct radiator dynamic loudspeakers arranged coaxially and the electrical networks. A. High frequency mechanism mounted in front of the low frequency mechanism. B. High frequency mechanism mounted congruently with the low frequency mechanism.

2.3 HORN LOUDSPEAKER

Large-scale reproduction of sound, involving several acoustical watts, is quite commonplace. Since high power audio frequency amplifiers are costly, it is logical to reduce the amplifier output to a minimum by the use of high-efficiency loudspeakers. At the present time, horn loudspeakers seem to be the only satisfactory high-efficiency system for large-scale sound reproduction. A horn loudspeaker consists of an electrically driven diaphragm coupled to a horn. The principal virtue of a horn resides in the possibility of presenting practically any value of acoustical resistance to the generator. This feature is extremely valuable for obtaining maximum overall efficiency in the design of the acoustical system. When a suitable combination of horns is employed, directional characteristics which are independent of the frequency, as well as practically any type of directional pattern, may be obtained. The combination of high efficiency and the possibility of any directional pattern makes the horn loudspeaker particularly suitable for large-scale reproduction.

A. Horn Loudspeaker Systems

Six typical horn loudspeakers are shown in Fig. 2.19. All of the horns are coupled to diaphragms of dynamic loudspeaker mechanisms.

The small horn loudspeaker shown in Fig. 2.19A is used to cover the high frequency portion of the audio frequency range.

The small horn loudspeaker with straight side walls is shown in Fig. 2.19B. The purpose of the straight side walls of the horn is to provide a more uniform directivity pattern. The directivity aspect of the straight-side-wall horn will be considered in Section 2.3F.

A small horn loudspeaker consisting of a multiplicity of horns, shown in Fig. 2.19C, is termed a cellular horn loudspeaker. The purpose of the cellular horn design is to provide a more uniform directivity pattern. The directivity aspect of the cellular horn will be considered in Section 2.3F.

A large straight-axis horn loudspeaker, sometimes referred to as a trumpet loudspeaker, is shown in Fig. 2.19D. This loudspeaker may be designed to cover a major portion of the audio frequency range.

A folded horn loudspeaker is shown in Fig. 2.19E. This loudspeaker

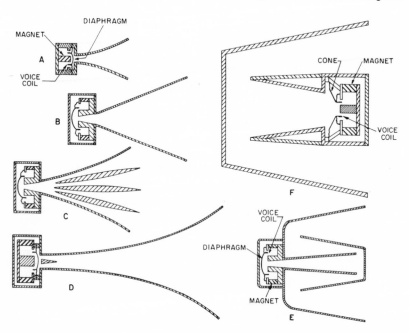

Fig. 2.19 Horn loudspeakers. A. High frequency horn loudspeaker. B. Straight side high frequency horn loudspeaker. C. Cellular high frequency horn loudspeaker. D. Trumpet horn loudspeaker. E. Folded horn loudspeaker. F. Low frequency folded horn loudspeaker.

covers the same frequency range as the trumpet loudspeaker of Fig. 2.19D since the length, the throat, and the mouth are the same. However, the loudspeaker of Fig. 2.19E is more compact and requires less space than the loudspeaker of Fig. 2.19D. The high frequency response is usually attentuated in a folded horn due to destructive interference incurred by the different path lengths of the sound traversing the bends. In order to eliminate destructive interference, the same phase should exist over any plane normal to the axis. This condition is practically satisfied provided the radial dimensions at the bends are a fraction of the wavelength. Wide-range reproduction of sound requires a large-mouth horn for efficient reproduction of low frequency sounds and small dimensions at the bends of a folded horn for efficient reproduction of high frequency sounds. Obviously, it is practically impossible to incorporate both of these features into a single folded horn. It is true that folded horns have been used for years, but, in general, the response at either or both the low or high frequency ranges has been attenuated.

A folded horn loudspeaker for the reproduction of the low frequency range is shown in Fig. 2.19F. Since this loudspeaker is used for the low frequency range, there is no problem of interference at the bends since the dimensions of the bends are a small fraction of the wavelength. As will be described in Section 2.3I, the low frequency horn loudspeaker of Fig. 2.19F is used in combination with one of the high frequency horn loudspeakers of Fig. 2.19B or Fig. 2.19C and thereby cover the entire audio frequency range.

B. Horn Loudspeaker Dynamic Mechanisms

The diaphragm, voice coil, magnet structure, and air chamber of a horn loudspeaker mechanism may be built in a wide variety of ways. The variations in path length from any part of the diaphragm to the horn throat should be less than a quarter wavelength in order to eliminate destructive interference in the air chamber. Two different methods for reducing interference in the air chamber are shown in Fig. 2.20B and C. These expedients are necessary for efficient reproduction at the high frequency portion of the audio range where the wavelength is relatively small. For the low frequency portion of the audio frequency range a large-throat horn may be coupled to a large diaphragm, as shown in Fig. 2.20A without incurring any loss due to interference, notwithstanding the large size, because the dimensions are small compared to the wavelength.

C. Diaphragms and Voice Coils

The diaphragms or cones of horn loudspeaker mechanisms are made of aluminum alloys, molded bakelite with various bases, molded styrol and

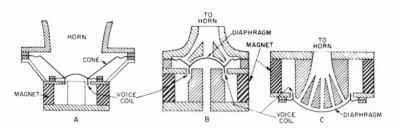

Fig. 2.20 Horn loudspeaker driving mechanisms. Mechanism A depicts a large diaphragm coupled to a large throat horn. Mechanisms B and C depict diaphragms and air chambers for coupling to a small throat horn.

other plastics, fiber, paper, and felted paper. Typical diaphragm shapes are shown in Fig. 2.20. Round, square, and ribbon wire conductors are used in voice-coil construction. Copper and aluminum conductors are used in the construction of the voice coil. (See Section 2.3H.)

D. Magnetic Field Structures

Permanent magnetic and electromagnetic field structures used in horn loudspeaker mechanisms are shown in Fig. 2.20. In general, it is customary to use higher flux densities in the gap in horn loudspeakers than in direct radiator loudspeakers. Soft iron may be used for the pole tips for flux densities up to 15,000 gausses. For flux densities from 20,000 to 23,000 gausses, an alloy of iron and cobalt is employed for the pole tip material in order to obtain these high densities with tolerable efficiency. (See Section 2.2C.)

E. Frequency Response Characteristics

The main virtue of a horn is to present an acoustic resistance to the diaphragm. Under these conditions efficiencies of the order of 50 percent may be obtained in horn loudspeakers. To obtain these efficiencies requires flux densities in the air gap of over 20,000 gausses. Furthermore, a horn design must be provided which will present an appropriate acoustic resistance to the diaphragm.

The cross-sectional area of an exponential horn in terms of the distance along the axis is given by:

$$S = S_0 \varepsilon^{mx} \tag{2.13}$$

where S = cross section, at the point x, in square centimeters,
$\quad\quad\quad S_0$ = cross section at the throat, in square centimeters,
$\quad\quad\quad \varepsilon$ = 2.718,
$\quad\quad\quad m$ = flare constant, and
$\quad\quad\quad x$ = distance from the throat, in centimeters.

The cutoff frequency of a horn due to flare is given by:

$$f_C = \frac{mc}{4\pi} \quad\quad\quad (2.14)$$

where f_C = cutoff frequency, in hertz,
$\quad\quad\quad c$ = velocity of sound, in centimeters per second, and
$\quad\quad\quad m$ = flare constant, in Equation (2.13).

Three exponential horns are shown in Fig. 2.21. The mouth termination is the same as that of a piston in an infinite baffle as depicted in Fig. 1.6. In Fig. 2.21A the radiation resistance is sufficiently high so that there is very little reflection of sound at the mouth of the horn. Therefore, the frequency response is relatively smooth. In Fig. 2.21B the radiation resistance at the mouth is $\frac{1}{16}$ that of Fig. 2.21A. As a consequence, the coupling to the air is poor. The result is a nonuniform frequency response characteristic. In Fig. 2.21C the mouth opening of the horn is the same as in Fig. 2.21B. However, the flare cutoff occurs at $D/\lambda = 0.4$. Below this frequency the response falls off quite rapidly.

In summary, the flare cutoff determines the lower limit of response of a

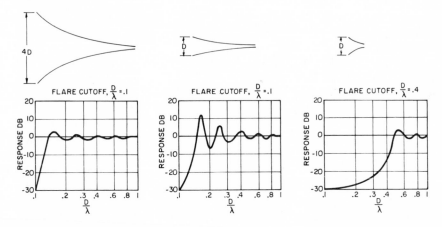

Fig. 2.21 Typical response frequency characteristics of horn loudspeaker for different mouth openings and different flare cutoffs as function of the ratio of the mouth diameter to the wavelength.

horn loudspeaker, and the ratio of the mouth dimension to the wavelength determines the smoothness of response as depicted by the illustrations of Fig. 2.21.

F. Directional Characteristics

The directional characteristics of a horn loudspeaker depend upon the shape, the mouth opening, and the frequency. This is an exceedingly complex subject; this section will consider only the major factors involved in the directivity of horns.

The directivity pattern of a typical exponential horn is shown in Fig. 2.22. The directivity becomes progressively sharper with increase of the frequency. This is the case for all single exponential horns. The mouth dimensions determine the directivity pattern up to $D = \lambda$. Above this frequency the flare of the horn plays the important role in the shape of the directivity pattern.

A horn with straight sides is shown in Fig. 2.23. The sharpest directivity occurs when the mouth dimension is equal to the wavelength. Below this frequency the directivity pattern becomes progressively broader with decrease of the frequency. Above the frequency where the mouth dimension is equal to twice the wavelength, the directivity is practically a constant over an angle determined by the angle of the side walls of the horn. Operation in this frequency range provides uniform directivity.

The use of a multicellular horn of Fig. 2.24 provides another means of obtaining a uniform directivity pattern in the frequency range above $D = 2\lambda$.

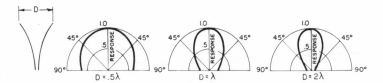

Fig. 2.22 Typical directional characteristics of exponential horn loudspeakers for the ratio of the diameter to the wavelength.

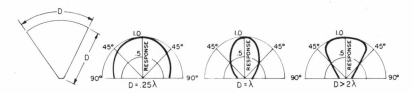

Fig. 2.23 Typical directional characteristics for a horn with straight side walls as a function of the dimension D to the wavelength.

As in the case of the straight sides, the sharpest directivity occurs for $D = \lambda$. Below this frequency range the directivity pattern becomes progressively broader.

G. Performance Considerations

A sectional view, electrical circuit, and mechanical network of a horn loudspeaker are shown in Fig. 2.25. The performance considerations in this section will involve the system depicted in Fig. 2.25.

The interaction of the current in the voice coil with the magnetic field produces a corresponding force which is transmitted to the cone. The force generated in the voice coil is given by:

$$f_M = Bli \qquad (2.15)$$

where f_M = force, in dynes,
   B = flux density in the air gap, in gausses,
   l = length of the voice coil conductor, in centimeters, and
   i = current in the voice coil in abamperes.

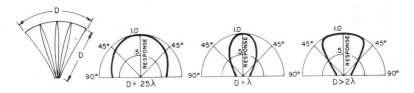

Fig. 2.24 Typical directional characteristics of a cellular horn loudspeaker as a function of the dimension D to the wavelength.

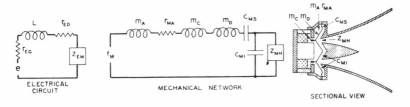

Fig. 2.25 Sectional view, voice coil electrical circuit and mechanical network of the vibrating system of a horn dynamic loudspeaker mechanism. In the electrical circuit: e = internal voltage of the generator. r_{EG} = internal electrical resistance of the generator. L = inductance of the voice coil. r_{ED} = damped electrical resistance of the voice coil. z_{EM} = motional electrical impedance. In the mechanical network: m_A and r_{MA} = mass and mechanical radiation resistance due to the air load on the back of the diaphragm. m_C and m_D = masses of the voice coil and diaphragm. C_{MS} and C_{M1} = compliances of the suspension and air chamber. z_{MH} = mechanical impedance at the throat of the horn. f_M = driving force.

The current in the voice coil can be determined from the electrical circuit of Fig. 2.25 as follows:

$$i = \frac{e}{r_{EG} + r_{ED} + j\omega L + z_{EM}} \tag{2.16}$$

where e = voltage of the generator, in abvolts,
   r_{EG} = electrical resistance of the generator, in abohms,
   r_{ED} = electrical resistance of the voice coil, in abohms,
   z_{EM} = electrical motional impedance due to the mechanical system, in abohms,
   L = inductance of the voice coil, in abfarads,
   $\omega = 2\pi f$, and
   f = frequency in hertz.

The electrical motional impedance is given by:

$$z_{EM} = \frac{(Bl)^2}{z_M} \tag{2.17}$$

where z_{EM} = electrical motional impedance, in abohms,
   B = flux density in the air gap, in gausses,
   l = length of the voice coil conductor, in centimeters, and
   z_M = mechanical impedance at f_M in Fig. 2.25 in mechanical ohms.

In the mechanical network, Fig. 2.25, the mechanical impedance, z_M, in mechanical ohms, at f_M is given by:

$$z_M = j\omega m_A + r_{MA} + j\omega m_C + j\omega m_D + \frac{1}{j\omega C_{MS}} + \frac{z_{MH}}{j\omega C_{M1} z_{MH} + 1} \tag{2.18}$$

where m_A = mass of the air load on the back of the diaphragm, in grams,
   m_C = mass of the voice coil, in grams,
   m_D = mass of the diaphragm, in grams,
   $\omega = 2\pi f$,
   f = frequency, in hertz,
   r_{MA} = mechanical resistance load on the back of the diaphragm, in mechanical ohms,
   C_{MS} = compliance of the suspension, in centimeters per dyne,
   C_{M1} = compliance of the air chamber, in centimeters per dyne,
   $z_{MH} = r_{MH} + jx_{MH}$ = mechanical impedance of the throat of the horn, in mechanical ohms,
   r_{MH} = mechanical resistance of the throat of the horn, in mechanical ohms, and
   x_{MH} = mechanical reactance of the throat of the horn, in mechanical ohms.

The velocity of the diaphragm of the horn loudspeaker is given by:

$$\dot{x} = \frac{f_M}{z_M} \qquad (2.19)$$

where $\dot{x}$ = velocity of the diaphragm.

H. Efficiency

The efficiency of a loudspeaker is the ratio of the sound power output to the electrical power input. The efficiency of a horn loudspeaker is given by:

$$\mu = \frac{B^2}{\left(\dfrac{\rho c A_D^2 \rho_c K_r}{m_c A_T}\right) 10^3 + B^2} \times 100 \qquad (2.20)$$

where μ = efficiency, in percent,
   B = flux density in the air gap, in gausses,
   A_D = area of the diaphragm, in square centimeters,
   ρ = density of air, in grams per cubic centimeter,
   c = velocity of sound, in centimeters per second,
   ρ_c = density of the conductor of the voice coil, in grams per cubic centimeter,
   K_r = resistivity of the voice coil conductor, in microhms per centimeter cube,
   m_C = mass of the voice, in grams, and
   A_T = area of the throat of the horn, in square centimeters.

In horn loudspeakers one of the objectives is to provide a high efficiency and thereby reduce the size of the amplifier for large-scale sound reproduction. Therefore, the flux density in the air gap is usually 15,000 to 22,000 gausses.

Copper and aluminum are used for the conductor in the construction of voice coils for horn loudspeakers. For copper the product $\rho_c K_r$ is 15.2 in grams per cubic centimeter times microhms per centimeter cube and for aluminum $\rho_c K_r$ is 7.6 in grams per cubic centimeter times microhms per centimeter cube. Equation (2.20) shows that considerable increase in efficiency may be obtained by the use of an aluminum coil. Therefore, in general, aluminum is used for the conductor in horn loudspeakers.

The efficiencies of commercial horn loudspeakers range from 20 to 50 percent.

I. Multiple Horn, Multiple Channel

Multiple horn loudspeakers each operating in different frequency ranges are used to provide uniform response and directivity and adequate sound power over the entire audio frequency range. The two most common "two-way" horn loudspeaker systems are shown in Fig. 2.26. In Fig. 2.26A a folded horn is used in the low frequency loudspeaker. In Fig. 2.26B a straight-axis horn is used in the low frequency loudspeaker. In order to minimize time delay and phase distortion due to a path-length difference between the low and high frequency horns, the effective lengths of the low and high frequency horns must be practically the same. The difference in path length in the system of Fig. 2.26A is made relatively small by employing a short folded horn coupled to a large-diameter dynamic loudspeaker mechanism of the type shown in Fig. 2.20A. A further reduction in path length may be obtained by the use of a straight-axis horn of Fig. 2.26B. In this case the length of the low and high frequency horns are practically the same. A cellular-horn-type high frequency loudspeaker is shown in Fig. 2.26A and a straight-side-type horn high frequency loudspeaker is shown in Fig. 2.26B. Either of the high frequency loudspeakers may be used with either of the low frequency loudspeakers. Both types provide a uniform directivity pattern. (See Section 2.3F.) The electrical network depicted in Fig. 2.26 is used to allocate the electrical input to the low and high frequency units in the appropriate frequency ranges. The loudspeakers of Fig. 2.22 are used for large-scale sound reproduction where high efficiency and large power outputs are required.

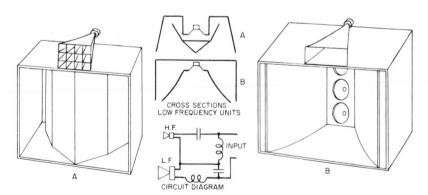

Fig. 2.26 Two channel, large scale, horn loudspeaker systems. A. A cellular horn high frequency loudspeaker and a folded horn loudspeaker. B. A straight side high frequency horn loudspeaker and a straight axis horn low frequency loudspeaker. The sectional views of the low frequency loudspeakers and the electrical circuit diagram are also depicted.

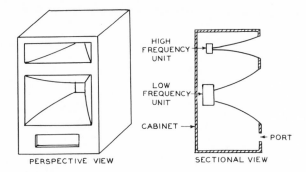

Fig. 2.27   Two channel horn loudspeaker consisting of straight axis low and high frequency horn units housed in a console cabinet.

A smaller version of the horn loudspeakers of Fig. 2.26 is shown in Fig. 2.27 and consists of straight-axis low and high frequency horns housed in a console cabinet. The electrical diagram is the same as that depicted in Fig. 2.26. A multicellular high frequency horn may be used instead of the high frequency horn with straight sides as shown in Fig. 2.27. Both types provide uniform directivity patterns. (See Section 2.3F.) The horn loudspeaker of Fig. 2.27 is used for monitoring in recording and broadcasting for medium-size sound reenforcement and for some home sound-reproducing systems.

Smaller versions of the low frequency loudspeaker shown in Figs. 2.26A and 2.19F, with the mouth operating into the corner of the room, have been used for sound reproduction in a residence. When the mouth operates into the corner of the room the radiation resistance is eight times that of the free-space condition. Therefore, a smaller mouth can be employed and still retain an adequate radiation resistance.

2.4 COMBINATION HORN AND DIRECT RADIATOR LOUDSPEAKERS

The combination horn and direct radiator loudspeakers consist of a large direct radiator loudspeaker for the reproduction of the low frequency portion of the audio frequency range and a small horn for the reproduction of the high frequency portion of the audio frequency range.

The combination horn and direct radiator loudspeaker shown in Fig. 2.28A consists of a direct radiator dynamic loudspeaker with a large cone, a simple straight-axis horn loudspeaker, and an electrical network for allocating the electrical input to the two loudspeakers to the appropriate frequency ranges.

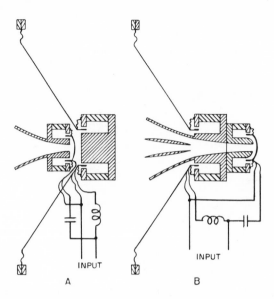

A   B

Fig. 2.28 Combination horn and direct radiator loudspeakers and the electrical networks. A. Small horn loudspeaker mounted in front of the cone of the direct radiator loudspeaker mechanism. B. The center poles of the low and high frequency mechanisms constitute a part of the cellular high frequency horn unit.

In Fig. 2.28B a part of the horn of the high frequency unit is formed by the pole pieces of the magnetic structures of the low and high frequency units. The remainder of the horn is of the cellular type which provide a uniform directivity pattern. (See Section 2.3F.)

Other combination horn and direct radiator loudspeakers are shown in Fig. 2.29. The loudspeakers shown in Fig. 2.29A and B consist of a direct radiator dynamic loudspeaker mounted in a completely enclosed cabinet. (See Section 2.2J.) The high frequency unit in Fig. 2.29A employs a straight axis horn to provide uniform directivity in the high frequency range. (See Section 2.3F.) The high frequency unit in Fig. 2.29B employs a cellular horn to provide uniform directivity in the high frequency range. (See Section 2.3F.) A combination horn and multiple direct radiator loudspeakers are shown in Fig. 2.29C. The high frequency horn is of the cellular type to provide a uniform directivity pattern in the high frequency range. A straight-axis horn could also be used to accomplish the same purpose. Four large direct radiator dynamic loudspeakers are used to reproduce the low frequency range. The loudspeaker shown in Fig. 2.29C is used for large-scale sound reproduction in which large acoustic outputs are required.

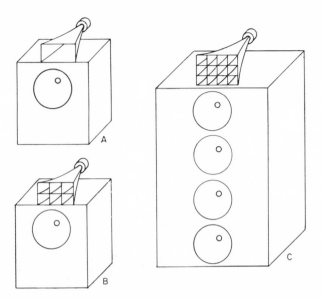

Fig. 2.29 Combination horn and direct radiator loudspeakers. A. Straight side horn high frequency loudspeaker and a single direct radiator low frequency loudspeaker mounted in a completely enclosed cabinet. B. Cellular horn high frequency loudspeaker and a single direct radiator low frequency loudspeaker mounted in a completely enclosed cabinet. C. A cellular horn loudspeaker and four direct radiator low frequency loudspeakers mounted in a completely enclosed cabinet.

2.5 LOUDSPEAKER PERFORMANCE REQUIREMENTS FOR HIGH FIDELITY SOUND REPRODUCTION

The term high fidelity sound reproduction is used to designate a superior order of performance. The characteristics involved in describing the performance of a loudspeaker are: frequency response, directivity, nonlinear distortion, transient response, impedance, efficiency, and sound power output. The purpose of the sections which follow are twofold, namely, to describe the performance characteristics of a loudspeaker and to establish specifications with respect to the characteristics in order to achieve a high order of excellence of performance.

A. Frequency Response

The frequency response of a loudspeaker is the sound pressure output on the axis as a function of the frequency.

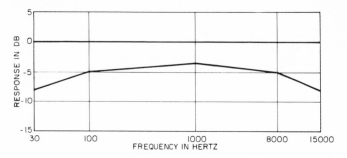

Fig. 2.30 The frequency response characteristic of a high-quality loudspeaker should fall between the depicted upper and lower limits of frequency response.

The loudspeaker should provide uniform response with respect to frequency over the frequency range from 30 to 15,000 hertz. For most high-quality applications the response should be contained within the limits of Fig. 2.30. As indicated in this chapter, there is no problem in achieving this performance in a high-quality loudspeaker.

B. Directivity

The directional characteristic of a loudspeaker is the response as a function of the angle with respect to some reference axis of the system. The directional patterns are usually depicted in polar coordinates.

If the directivity pattern varies with frequency, frequency discrimination will occur for observation and listening points removed from the axis. Uniform directivity is particularly important in stereophonic sound reproduction in order to provide realistic auditory perspective.

Limits on the directivity pattern of a loudspeaker over the response range of 30 to 15,000 Hz are depicted in Fig. 2.31. The polar curves should fall within the amplitude limits shown for an angle of at least 45°. For really high-quality performance, the polar curves should fall within the amplitude limits shown for an angle of 90°. The limits on the directivity pattern of Fig. 2.31 apply to loudspeakers for reproduction of sound by a consumer in a residence and do not necessarily apply to applications in a theater, an auditorium, etc.

C. Nonlinear Distortion

The nonlinear distortion frequency characteristic of a loudspeaker is the total nonlinear distortion as a function of the frequency.

The major result of nonlinearity in the elements of the vibrating system of a direct radiator loudspeaker is the production of harmonics and subharmonics. Two major contributors to nonlinear distortion in dynamic loudspeaker

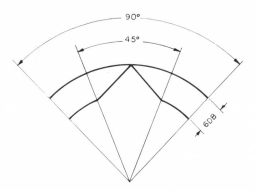

Fig. 2.31   The directional characteristic of a high quality loudspeaker should fall between the limits depicted over the frequency range of 30 to 15,000 hertz.

mechanisms are the driving and suspension elements. These elements are constant for small and moderate amplitudes but depart from constancy for large excursions of the cone or diaphragm. In the low frequency range the amplitude of the cone in the direct radiator loudspeaker must be inversely proportional to the square of the frequency, and the amplitude of the diaphragm in the horn loudspeaker must be inversely proportional to the frequency in order to maintain constant output with respect to frequency. Thus amplitude increases rapidly with a decrease of the frequency. For the large amplitudes there is considerable departure from linearity in the operation of the driving and suspension systems.

A typical nonlinear distortion frequency characteristic of a direct radiator dynamic loudspeaker with a nonlinear driving and suspension system is depicted by the curve *A* of Fig. 2.32. The distortion decreases as the power input decreases.

Distortion due to the nonlinear driving system can be eliminated by a voice coil design as depicted in Fig. 2.7C. In this configuration the flux-turn product will be a constant for very large amplitudes. The nonlinear distortion, employing the voice coil of Fig. 2.7C, is shown as curve *B* of Fig. 2.32. The nonlinear distortion of curve *A* of Fig. 2.32 is for the voice coil of Fig. 2.7B.

Distortion due to a nonlinear suspension system can be reduced with a low resonant frequency loudspeaker mechanism and a completely enclosed cabinet in which the the cabinet is the controlling compliance. The distortion is reduced to curve *C* of Fig. 2.32 by the use of the voice coil of Fig. 2.7C and the use of a low resonant loudspeaker mechanism and the completely enclosed cabinet of Fig. 2.13C.

A further reduction in nonlinear distortion can be obtained by the use of the complementary voice coil–suspension design of Fig. 2.8. By means of this system the nonlinear distortion is reduced to curve *D* of Fig. 2.32.

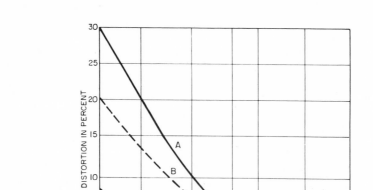

Fig. 2.32 Typical nonlinear distortion frequency characteristics of direct radiator dynamic loudspeakers. Curve *A*. Voice coil of Fig. 2.7B and the suspension of Fig. 2.5A. Curve *B*. Voice coil of Fig. 2.7C and the suspension of Fig. 2.5B. Curve *C*. The voice coil of 2.7C and the suspension of 2.5C with the mechanism housed in a completely enclosed cabinet. Curve *D*. The complementary voice coil-suspension design with the mechanism housed in a completely enclosed cabinet.

The cones and diaphragms of loudspeakers should be designed so that the operation falls within the limits of Hooke's law for the material and construction. This leads to a relatively heavy cone or diaphragm. This poses no problem except for a reduction in sensitivity. The subject of sensitivity will be discussed in a later section. When a heavy cone or diaphragm is employed, the distortion in the frequency range above 200 hertz is exceedingly low.

D. Transient Response

The transient response of a loudspeaker refers to the faithfulness of response of a loudspeaker to a sudden change in the electrical input.

For the most part the sounds of speech and music are of a transient rather than a steady state character. Therefore, practically all sounds which are reproduced in a sound-reproducing system are of a transient nature. As a consequence, the transient response of a loudspeaker must be considered an important factor in the performance of the loudspeaker.

The test employed to depict the transient response of a loudspeaker is the application of a tone burst. The frequency response characteristic of a loudspeaker and the tone burst response are shown in Fig. 2.33. The tone burst response is quite faithful in the smooth part of the frequency response characteristic. However, there is considerable deviation of the tone burst output of the loudspeaker as compared to the tone burst input at the peak and dip in the frequency response characteristic. In a properly designed loudspeaker, if the frequency response is smooth the transient response will be faithful.

E. Electrical Impedance

The frequency electrical impedance characteristic of a loudspeaker is the electrical impedance of the loudspeaker as a function of the frequency.

In the dynamic-type loudspeaker the variation of the impedance with frequency is relatively small and there is normally no problem in the transfer of power from the amplifier to the loudspeaker because of the electrical impedance characteristic.

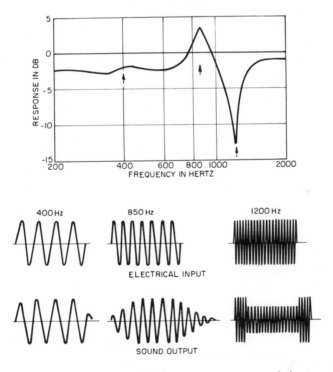

Fig. 2.33 The relationship between the frequency response and the tone burst at 400, 850 and 1200 hertz.

F. Efficiency

The efficiency of a loudspeaker is the ratio of the sound power output to the electrical power input.

For reproduction of sound in small rooms such as in the home, monitoring rooms, offices, etc., the low efficiency exhibited by the general run of direct radiator dynamic loudspeaker is of small concern because the maximum power levels needed can be obtained with relatively small and modest power amplifiers. However, for large-scale sound reproduction, as for example, sound-reinforcing systems in large auditoriums, efficiency is a consideration if the amplifier requirements are not to become ridiculously large.

G. Maximum Sound Power Output

The maximum sound power output of a loudspeaker is determined by the tolerable or allowable limits upon the distortion or other overload factors such as excessive heating of the voice coil.

2.6 SOUND POWER EMITTED BY A LOUDSPEAKER

The sound power emitted by a point source or by a nondirectional (omnidirectional) loudspeaker is given by:

$$P_{AN} = \frac{4\pi r^2 p^2}{\rho c} \tag{2.21}$$

where P_{AN} = sound power, in ergs per second,
p = sound pressure, in microbars,
r = distance, in centimeters,
ρ = density of air, in grams per cubic centimeter, and
c = velocity of sound, in centimeters per second.

If the sound source or loudspeaker is directional, that is, if the sound emitted varies with the direction with respect to some axis in the system, then the total sound power from the geometry of Fig. 2.34 is given by:

$$P_{AD} = \frac{r^2}{\rho c} \int_0^{2\pi} \int_0^{\pi} p^2(\theta, \psi, r) \sin \theta \, d\theta \, d\psi \tag{2.22}$$

where P_{AD} = sound power, in ergs per second,
$p(\theta, \psi, r)$ = sound pressure, in microbars, at a distance r, in centimeters, and the angles θ and ψ, and
θ and ψ = angular polar coordinates of the system. The loudspeaker axis coincides with the x axis.

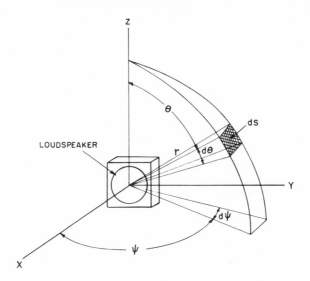

Fig. 2.34   The geometry for obtaining the total sound power output of a loudspeaker.

2.7 LOUDSPEAKER DIRECTIVITY FACTOR AND INDEX

The directivity factor of a loudspeaker is the ratio of the sound power which would be radiated if the free-space axial sound pressure were constant over 4π solid angles to the actual sound power radiated. The directivity factor is given by:

$$Q = \frac{P_{AN}}{P_{AD}} \qquad (2.23)$$

where
$$Q = \text{directivity factor,}$$
$$P_{AN} = \text{Equation (2.21), and}$$
$$P_{AD} = \text{Equation (2.22).}$$

The directivity factor can be computed from the directional characteristics of the loudspeaker employing Equation 2.23.

The directivity index of a loudspeaker can be obtained from the directivity factor by use of the following equation:

$$DI_{DB} = 10 \log_{10} Q \qquad (2.24)$$

REFERENCES

Hilliard, John K. "An Improved Theater Type Loudspeaker," *Journal of the Audio Engineering Society*, Vol. 17, No. 5, p. 512, 1969.

Olson, Harry F. *Acoustical Engineering*, Van Nostrand Reinhold, New York, 1957. Treats the following subjects not considered in this chapter, namely, the extended theory of direct radiator and horn loudspeakers, and electrostatic, corner horn, throttled air flow, and ionic loudspeakers, along with ninety references.

Olson, Harry F. *Solutions of Engineering Problems by Dynamical Analogies*, Van Nostrand Reinhold, New York, 1966. Provides the means for deriving and solving the dynamical analogies considered in this chapter.

Olson, Harry F. "Direct Radiator Loudspeaker Enclosures, Reissue," *Journal of the Audio Engineering Society*, Vol. 17, No. 1, p. 22, 1969.

Olson, Harry F. "Analysis of the Effects of Nonlinear Elements Upon the Performance of a Back Enclosed Loudspeaker Mechanism," *Journal of the Audio Engineering Society*, Vol. 10, No. 2, p. 156, 1962.

Small, Richard H. "Constant-Voltage Crossover Network Design," *Journal of the Audio Engineering Society*, Vol. 19, No. 1, p. 12, 1971. Although the crossover networks depicted in this book are of the parallel type, the series crossover networks may also be used, as described in this reference.

Tremaine, Howard M. *Audio Cyclopedia*, Howard W. Sams, Indianapolis, 1969. (See also numerous articles on loudspeakers in the *Journal of the Audio Engineering Society*).

3
Earphones

3.1 INTRODUCTION

An earphone is an electroacoustic transducer closely coupled acoustically to the ear and generating a sound pressure at the ear in which the resultant acoustic waveform is essentially equivalent to the electrical input. The most common earphones in use today are the magnetic, crystal, dynamic, and electrostatic. The purpose of this chapter is to describe the constructional features and performance characteristics of earphones.

3.2 MAGNETIC EARPHONES

Magnetic earphones are electroacoustic transducers in which the forces which actuate the vibrating system are generated by the interaction of magnetic fields in magnetic elements.

A. Bipolar Earphones

A sectional view, mechanical network, electrical circuit, and frequency response characteristic of a bipolar earphone are shown in Fig. 3.1. The

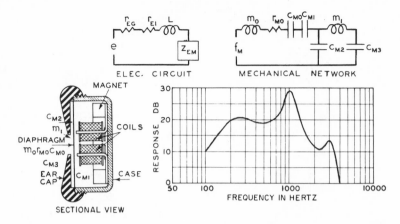

Fig. 3.1   Sectional view, mechanical network, electrical circuit and frequency response characteristics of a bipolar earphone. In the mechanical network, f_M = the mechanical driving force. m_0, r_{MO}, and C_{MO} = mass, mechanical resistance, and compliance of the diaphragm. C_{M1} = compliance due to the air in the case. C_{M2} = compliance of the air space between the diaphragm and cover. m_1 = mass of the air in the aperture in the cover. C_{M3} = compliance of the ear cavity. In the electrical circuit, z_{EM} = motional electrical impedance. L and r_{E1} = damped inductance and electrical resistance of the coils. r_{EG} = electrical resistance of the electrical generator. e = voltage of the electrical generator. The graph shows the pressure frequency response characteristic.

bipolar earphone shown in Fig. 3.1 consists of a steel diaphragm spaced a small distance from the pole pieces, which are wound with insulated wire. A permanent magnet in contact with the pole pieces supplies the steady magnetic field through the pole pieces to the diaphragm. An alternating current in the coil produces an alternating flux superimposed upon the steady magnetic flux. The resultant force acting upon the diaphragm produces a motion of the diaphragm that corresponds to the input. The motion of the diaphragm produces a sound pressure at the ear which corresponds to the electrical current input. The resonant frequency of the diaphragm is usually about 1000 hertz. As a consequence, there is a peak in the response at 1000 hertz as shown by the frequency response characteristic of Fig. 3.1. Below the resonant frequency, the pressure response is maintained until the leak between the cap and the ear reduces the pressure response. Above the resonant frequency, the response falls off quite rapidly, with a very small output above 4000 hertz. Therefore, the bipolar telephone receiver is not suitable for high-quality applications. The use is confined to limited-frequency-range communication and test applications. The electrical circuit of Fig. 3.1 shows that the electrical impedance of the earphone is in series with the voltage of the electrical driving generator and the generator electrical resistance.

A sectional view, mechanical network, electrical circuit, and frequency response characteristic of a small version of the bipolar earphone are shown in Fig. 3.2. This earphone is equipped with an earpiece for insertion in the outer ear canal. The main application for this earphone is in hearing aids. The action is similar to the larger version of the bipolar telephone receiver. Since the diaphragm is smaller than the earphone of Fig. 3.1, the resonant frequency of the diaphragm occurs at 3500 hertz. Therefore, the frequency response characteristic is smooth up to 4000 hertz.

B. Magnetic Armature Earphone

A sectional view, mechanical network, electrical network, and frequency response characteristic of the magnetic armature earphone are shown in Fig. 3.3. The ring-type magnetic armature earphone shown in Fig. 3.3 consists of a driving armature of an iron–cobalt alloy coupled to a dome-shaped diaphragm of a light aluminum alloy. The steady magnetic flux is supplied by a permanent magnet. An alternating current in the coil produces a corresponding alternating flux that is interimposed upon the steady flux. A corresponding resultant force is developed in the armature and transferred

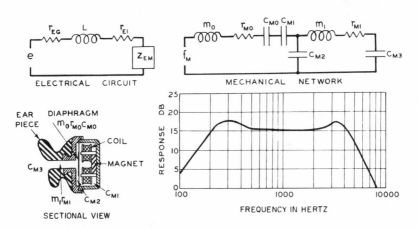

Fig. 3.2 Sectional view, mechanical network, electrical circuit, and frequency response characteristic of an insert-type earphone. In the mechanical network, f_M = mechanical driving force. m_O, r_{MO}, and C_{MO} = mass, mechanical resistance, and compliance of the diaphragm. C_{M1} = compliance due to the air in the case. C_{M2} = compliance of the air space between the diaphragm and the cover. m_1 and r_{M1} = mass and mechanical resistance of the tube. C_{M3} = compliance of the ear cavity. In the electrical circuit, z_{EM} = electrical motional impedance. L and r_{E1} = damped inductance and electrical resistance of the coil. r_{EG} = electrical resistance of the coils. e = voltage of the electrical generator. The graph depicts the frequency response characteristic.

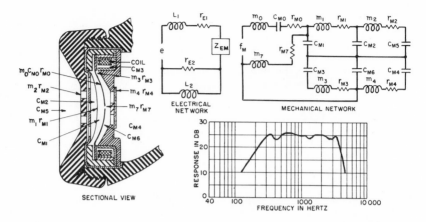

Fig. 3.3 Sectional view, mechanical network, electrical circuit, and the frequency response characteristic of a ring armature earphone. In the mechanical network, $f_M =$ mechanical driving force. m_0, r_{MO}, and C_{MO} = mass, mechanical resistance and compliance of the diaphragm. m_1 and r_{M1} = mass and mechanical resistance of the thin membrane. C_{M1} = compliance of the air space between the diaphragm and the membrane. m_2 and r_{M2} = mass and mechanical resistance of the holes in the ear cap. m_3 and r_{M3} = mass and mechanical resistance of the air gap aperture. C_{M2} = compliance of the air volume between the ear cap and the thin membrane. C_{M3} = compliance of the air volume in the coil space. m_4 and r_{M4} = mass and mechanical resistance of the control mechanical resistance. C_{M4} = compliance of the air volume in the handle. C_{M5} = compliance of the air cavity between the ear cap and the ear. C_{M6} = compliance of the air cavity between the diaphragm and back enclosure. m_7 and r_{M7} = mass and mechanical resistance of the small hole in the diaphragm. In the electrical network, z_{EM} = motional electrical impedance. L_1 and r_{E1} = damped inductance and electrical resistance of the coil. L_2 and r_{E2} = inductance and electrical resistance due to eddy currents. e = the voltage of the electrical generator. The graph shows the frequency response feeding a cavity.

to the diaphragm. The performance of the system can be obtained from the mechanical network. The electrical network of Fig. 3.3 shows the elements that determine the actuating current in the coil in terms of the electrical generator. The frequency response characteristic of the magnetic armature earphone shown in Fig. 3.3 is smooth over the frequency range 200 to 4000 hertz. The earphone shown in Fig. 3.3 is used in the telephone handset of Fig. 7.2.

3.3 CRYSTAL EARPHONE

A sectional view, mechanical network, electrical circuit, and frequency response characteristic of a crystal earphone are shown in Fig. 3.4. The crystal earphone shown in Fig. 3.4 consists of a light conical paper diaphragm

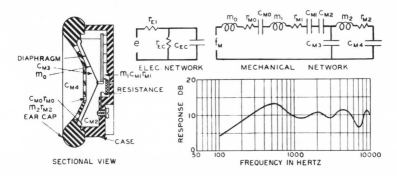

Fig. 3.4 Sectional view, mechanical network, electrical network, and frequency response characteristic of a crystal earphone. In the mechanical network, f_M = mechanical driving force. m_0, r_{M0}, and C_{M0} = mass, mechanical resistance, and compliance of the diaphragm. m_1, r_{M1}, and C_{M1} = mass, mechanical resistance, and compliance of the crystal. C_{M2} = compliance due to the air in the case. C_{M3} = compliance of the air space between diaphragm and cover. m_2 and r_{M2} = mass and mechanical resistance of the holes in the cover. C_{M4} = compliance of the ear cavity. In the electrical network, C_{EC} and r_{EC} = electrical capacitance and electrical resistance of the crystal. r_{E1} = electrical resistance of the series resistor. e = voltage of the electrical generator. The graph shows the frequency response characteristic.

coupled to a Rochelle salt crystal. Three corners of the bender type, bimorph crystal are fastened to the case. The fourth corner of the crystal is connected to the diaphragm. A ceramic element has also been used to drive the diaphragm. (See Section 4.3C.) The application of a voltage to the crystal produces a corresponding deformation that is transferred to the diaphragm. The amplitude of the diaphragm is proportioned to the voltage applied to the crystal. The performance of the system can be determined from the mechanical network. The frequency response characteristic of the crystal earphone is shown in Fig. 3.4. Since the crystal earphone is a stiffness-controlled system, the seal between the ear and the earphone must be perfect in order to maintain the low frequency response. The electrical network of the crystal earphone in Fig. 3.4 shows that the crystal is an electrical capacitance. The electrical capacitance of the earphone is in series with the voltage of the electrical driving generator and the generator electrical resistance. To smooth out the electrical impedance presented to the electrical driving generator with respect to frequency, a resistor is shunted across the crystal. Excellent high frequency response and light weight are features of the crystal earphone.

A small version of the crystal earphone equipped with an earpiece of a type similar to the small version of the bipolar earphone of Fig. 3.2 has also been commercialized. Its performance is similar to that of the larger version of the crystal earphone shown in Fig. 3.4.

3.4 DYNAMIC EARPHONE

A sectional view, mechanical network, electrical circuit, and frequency response characteristic of the dynamic earphone are shown in Fig. 3.5. The dynamic earphone in the figure consists of a light conical diaphragm of paper or plastic connected to a voice coil located in a magnetic field. The vibrating system is similar to a small dynamic loudspeaker mechanism. (See Section 2.2.) The mechanical network of Fig. 3.5 shows that the pressure response in the low part of the audio frequency range is maintained by the use of a low resonant frequency and a soft liquid-filled plastic ear cap to reduce the leakage between the ear cap and the other ear. The mechanical network also shows that the use of a small light-weight diaphragm and voice coil insures that the response will be maintained in the high frequency part of the audio frequency range. Employing a suitable design, the dynamic earphone will reproduce the entire audio frequency range with good fidelity as

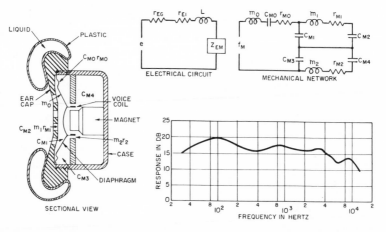

Fig. 3.5 Sectional view, mechanical network, electrical circuit and frequency response characteristic of a dynamic earphone. In the mechanical network: f_M = driving force. $f_M = Bli$. B = flux density in the air gap. l = length of the conductor in the voice coil. i = current in the voice coil. m_0 = mass of the diaphragm and voice coil. r_{MO} and C_{MO} = mechanical resistance and compliance of the diaphragm suspension system. m_1 and r_{M1} = mass and mechanical resistance of the holes in the ear cap. C_{M1} = compliance of the air in the cavity between the diaphragm and the ear cap. C_{M3} = compliance of the cavity behind the diaphragm. m_2 and r_{M2} = mass and mechanical resistance of the air between the voice coil and the pole. C_{M4} = compliance of the air in the case. C_{M2} = compliance of the ear cavity. In the electrical circuit: e = voltage of the electrical generator. r_{EG} = electrical resistance of the electrical generator. r_{E1} and L = electrical resistance and inductance of the voice coil. z_{EM} = motional electrical impedance. The graph shows the frequency response characteristic.

shown by the frequency response characteristic of Fig. 3.5. The electrical circuit of Fig. 3.5 shows that the electrical impedance of the earphone is in series with the voltage of the electrical driving generator and the generator electrical resistance.

3.5 ELECTROSTATIC EARPHONE

A sectional view, mechanical network, electrical network, and frequency response characteristic of the electrostatic earphone are shown in Fig. 3.6. The electrostatic earphone shown in Fig. 3.6 consists of a metal-coated thin plastic diaphragm located between perforated plates. Referring to the electrical circuit and the mechanical network of Fig. 3.6, the force applied to the diaphragm is proportional to the product of the signal voltage and the polarizing voltage. Therefore, the motion of the diaphragm and the sound pressure delivered to the ear correspond to the electrical signal input to the

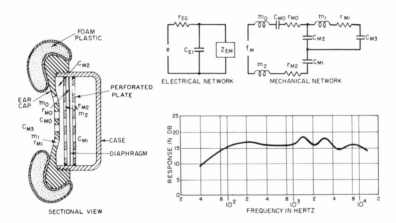

Fig. 3.6 Sectional view, mechanical network, electrical network and frequency response characteristic of an electrostatic earphone. In the mechanical circuit: f_M = driving force. $f_M = e_0 e A / 4 a^2$. e_0 = polarizing voltage. e = generator voltage. A = area of the diaphragm. a = spacing between the diaphragm and the back plate. m_0, C_{MO} and r_{MO} = mass, compliance and mechanical resistance of the diaphragm. m_1 and r_{M1} = mass and mechanical resistance of the holes in the ear cap. C_{M2} = compliance of the air in the cavity between the diaphragm and the ear cap. m_2 and r_{M2} = mass and mechanical resistance of the holes in perforated back plate. C_{M1} = compliance of the air in the case. C_{M3} = compliance of the ear cavity. In the electrical network: e = voltage of the electrical generator. r_{EG} = electrical resistance of the electrical generator. C_{E1} = electrical capacitance of the earphone. z_{EM} = motional electrical impedance. The graph shows the frequency response characteristic.

earphone. Some plastics exhibit electret properties, that is, the diaphragm is self-polarized.

The electrical circuit for the self-polarized electrostatic earphone is shown in Fig. 3.7A. A higher polarizing voltage and greater sensitivity may be obtained by the use of the self-electrical-polarizing system of Fig. 3.7B, in which the audio signal is stepped up in voltage and rectified to supply a high voltage bias. There may be some variation in voltage as the audio signal varies, thus introducing a variation in sensitivity. This can be overcome by the use of the system shown in Fig. 3.7C, in which the high voltage obtained from the ac mains is rectified to thereby supply a steady high voltage bias. A soft plastic ear cap reduces the leakage between the ear and the earphone and thereby maintains the pressure response in the low frequency range. If a suitable design is employed, the electrostatic earphone will reproduce the entire audio frequency range as shown by the frequency response characteristic of Fig. 3.6.

3.6 BONE CONDUCTION EARPHONE

A sectional view, mechanical network, electrical circuit, and response frequency characteristic of a bone conduction earphone are shown in Fig. 3.8. In certain types of deafness, the middle ear, which consists of a series of bones that conduct the sound from the ear drum to the inner ear, is damaged, while the inner ear, which consists of nerves, is normal. (See Section 17.2.) Under these conditions, sound may be transmitted through the bones of the head to the inner ear by means of the bone conduction earphone of Fig. 3.8.

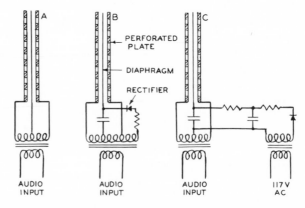

Fig. 3.7   Electrical systems for electrostatic earphones. A. Self polarized. B. Audio signal polarized. C. Power mains polarized.

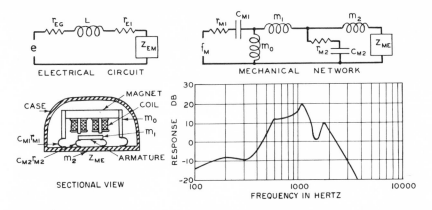

Fig. 3.8   Sectional view, mechanical network, electrical circuit, and frequency response characteristic of a bone conduction receiver. In the mechanical network, m_0 = mass of the coil and magnetic structure. m_1 = mass of the armature. C_{M1} and r_{M1} = compliance and the mechanical resistance connecting the coil and magnetic structure to the case. C_{M2} and r_{M2} = compliance and mechanical resistance connecting the armature to the case. m_2 = mass of the case. z_{ME} = mechanical impedance of the mastoid bone. f_M = the mechanical driving force. In the electrical circuit, z_{EM} = electrical motional impedance. L and r_{E1} = damped inductance and electrical resistance. r_{EG} = electrical resistance of the generator. e = voltage of the electrical generator. The graph depicts the frequency characteristic developed on an artificial mastoid.

The face of the bone conduction receiver is placed against the mastoid bone behind the ear. The mechanical network of Fig. 3.8 depicts a multiresonant vibrating system. The use of a multiresonant vibrating system makes it possible to deliver a relatively large driving force to the mastoid bone over a relatively wide frequency range. The frequency response characteristic of Fig. 3.8 shows that relatively good response is obtained over the useful audio range for intelligible speech.

3.7 EARPHONE PERFORMANCE REQUIREMENTS FOR HIGH FIDELITY SOUND REPRODUCTION

The earphones shown in Figs. 3.1, 3.2, 3.3, and 3.8 are not intended to provide uniform response over the entire audio frequency range, because for the applications of those earphones such a wide frequency range is not required. However, the earphones shown in Figs. 3.5 and 3.6 can provide high-quality performance. Since the amplitude of the diaphragm is small for very high sound levels delivered to the ear, there is no problem in achieving low nonlinear distortion and adequate power-handling capacity. The major

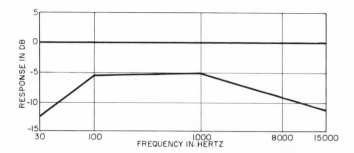

Fig. 3.9 The frequency response characteristic of a high-quality earphone should fall between these depicted upper and lower limits of frequency response.

concern is the frequency response characteristic. The response obtained on the artificial ear of Section 13.8 and Fig. 13.12 should provide uniform response over the frequency range of 30 to 15,000 hertz. For most high-quality applications the response should be contained within the limits shown in Fig. 3.9.

The earphones described in this chapter may be operated from a monaural or a binaural reproducer. (See Sections 6.2 and 6.3.) The earphones may also be operated from a monophonic or stereophonic reproducer. In the monaural or monophonic mode the same signal is applied to both earphones. In the binaural or stereophonic mode the two earphones are each connected to a channel.

Earphones attached to a headband are often referred to as a headset or headphones.

REFERENCES

Bubbers, John J. "Some Design Considerations for Electrostatic Earphones," *Journal of the Audio Engineering Society*, Vol. 19, No. 3, p. 206, 1971.

Mott, E. E. and Miner, R. C. "The Ring Armature Telephone Receiver," *Bell System Technical Journal*, Vol. 30, No. 2, p. 110, 1951.

Olson, Harry F. *Acoustical Engineering*, Van Nostrand Reinhold, New York, 1957.

Olson, Harry F. *Solutions of Engineering Problems by Dynamical Analogies*, Van Nostrand Reinhold, New York, 1966. Provides the means for deriving and solving the dynamical analogies considered in this chapter.

Selstead, Walter T. "The Electrostatic Earphone," *Journal of the Audio Engineering Society*, Vol. 9, No. 2, p. 145, 1961.

Sessler, G. M. and West, J. E. "Condenser Earphones with Solid Dielectric," *Journal of the Audio Engineering Society*, Vol. 10, No. 3, p. 212, 1962.

(See also numerous articles on earphones in the *Journal of the Audio Engineering Society*).

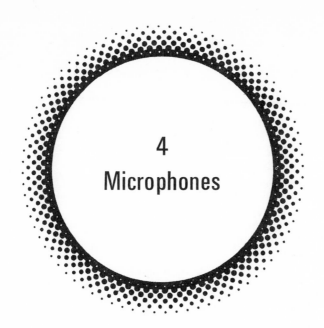

4
Microphones

4.1 INTRODUCTION

A microphone is an electroacoustic transducer actuated by power in an acoustical system and delivering power to an electrical system, the waveform in the electrical system being substantially equivalent to that in the acoustical system. A pressure microphone is a microphone in which the electrical response corresponds to the sound pressure in the sound wave. A velocity or pressure gradient microphone is a microphone in which the electrical response corresponds to the particle velocity or pressure gradient in a sound wave. All microphones in use today may be classed as pressure, velocity, pressure gradient, and combination pressure and velocity or pressure gradient. Microphones may also be classified with respect to the directional characteristic, as for example, nondirectional or omnidirectional, bidirectional, and unidirectional. The purpose of this chapter is to consider microphones from the standpoints of both the response and the directivity.

4.2 DIRECTIONAL PROPERTIES

A. Directional Characteristics

The directional characteristic of a microphone is the response as a function of the direction of the impinging sound wave with respect to some reference axis of the microphone.

The directional response of a nondirectional or an omnidirectional microphone is independent of the direction of the impinging sound wave. The directional characteristic of a nondirectional microphone is shown in Fig. 4.1A.

The directional response of a bidirectional microphone is given by:

$$R_D = \cos \theta \qquad (4.1)$$

where R_D = response for the angle θ, and
θ = angle between the direction of the impinging sound wave and the major axis of the microphone.

The directional characteristic of a bidirectional microphone is shown in Fig. 4.1B.

The directional response of a unidirectional microphone with a cardioid pattern is given by:

$$R_D = \frac{1 + \cos \theta}{2} \qquad (4.2)$$

where R_D = response for the angle θ, and
θ = angle between the direction of the impinging sound wave and the major axis of the microphone.

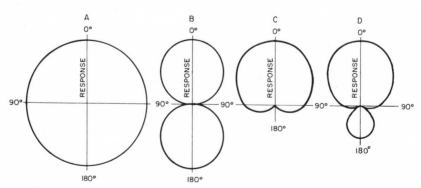

Fig. 4.1 Directional characteristics of microphones. A. Nondirectional. B. Bidirectional. C. Cardioid. D. Limacon of highest directivity.

The directional characteristic of a unidirectional microphone with cardioid pattern is shown in Fig. 4.1C.

The directional response of a unidirectional microphone with a limacon pattern of the highest directivity is given by:

$$R_D = \tfrac{1}{4} + \tfrac{3}{4} \cos \theta \qquad (4.3)$$

where R_D = response for the angle θ, and
θ = angle between the direction of the impinging sound wave and the major axis of the microphone.

The directional characteristic of a unidirectional microphone with a limacon pattern of the highest directivity is shown in Fig. 4.1D.

The directional characteristic of a bidirectional gradient microphone of the order n is given by:

$$R_D = \cos^n \theta \qquad (4.4)$$

where R_D = response for the angle θ,
θ = angle between the direction of the impinging sound wave and the major axis of the microphone, and
n = order of the gradient.

In this case, the directional characteristics are bidirectional cosine functions, the power of the cosine being the order of the gradient. The directional characteristics of bidirectional gradient microphones of the order of zero, one, two, three, and four are shown in Fig. 4.2.

The directional characteristic of a unidirectional microphone of the order

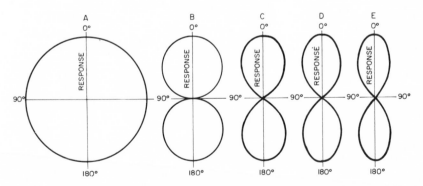

Fig. 4.2   Directional characteristics of bidirectional microphones of different orders. A. Order zero. B. Order one. C. Order two. D. Order three. E. Order four.

n' employing the cardioid pattern as the base is given by:

$$R_D = \frac{(1 + \cos \theta) \cos^{n'-1} \theta}{2}$$ (4.5)

where R_D = response for the angle θ,
      θ = angle between the direction of the impinging sound wave and the axis of the microphone, and
      n' = order of the gradient.

The directional characteristics where n' is a positive integer are undirectional directivity patterns. The directional characteristics of unidirectional microphones of order one, two, three, and four are shown in Fig. 4.3.

B. Directional Efficiency, Directivity Factor, and Directional Gain or Directivity Index

The ratio of energy response of a directional microphone to the energy response of a nondirectional microphone, all directions being equally probable, is termed the directional efficiency. The directional efficiency is given as follows:

$$D.E. = \frac{1}{4\pi} \int_0^{4\pi} f^2(\psi) \, d\Omega_\psi$$ (4.6)

where $D.E.$ = directional efficiency,
      $f(\psi)$ = ratio of the voltage output for incidence at the angle ψ to that for $\psi = 0$, and
      $d\Omega_\psi$ = element of solid angle at the angle ψ.

The directional efficiency of a microphone is a measure of the energy response to reverberation and undesirable noise.

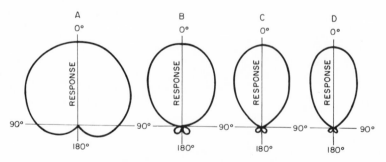

Fig. 4.3 Directional characteristics of unidirectional microphones of different orders. A. Order one. B. Order two. C. Order three. D. Order four.

The directivity factor is the reciprocal of the directional efficiency and is given by:

$$Q = \frac{1}{D.E.} \tag{4.7}$$

where $\qquad\qquad Q$ = directivity factor.

For the same ratio of signal to random noise, reverberation, etc., the directional microphone may be operated at $\sqrt{Q}$ times the distance of a nondirectional microphone. The directional gain or directivity index in terms of the directivity factor is given by Equation (2.24).

The directional efficiency, the directivity factor, and the directivity index for various directional microphones are shown in Table 4.1.

TABLE 4.1

The Directional Efficiency, Directivity Factor, and Directivity Index for Various Microphones

Directional Characteristic	Directional Efficiency	Directivity Factor	Directivity Index
$R_D = 1$	1	1	0
$R_D = \cos \theta$	0.33	3	4.8
$R_D = \cos^2 \theta$	0.20	5	7.0
$R_D = \cos^3 \theta$	0.14	7	8.5
$R_D = \cos^4 \theta$	0.11	9	9.5
$R_D = \frac{1}{2} + \frac{1}{2} \cos \theta$	0.33	3	4.8
$R_D = \frac{1}{4} + \frac{3}{4} \cos \theta$	0.25	4	6.0
$R_D = (\frac{1}{2} + \frac{1}{2} \cos) \cos \theta$	0.13	7.5	8.8
$R_D = (\frac{1}{2} + \frac{1}{2} \cos) \cos^2 \theta$	0.086	11.6	10.6
$R_D = (\frac{1}{2} + \frac{1}{2} \cos) \cos^3 \theta$	0.064	15.7	12.0

4.3 PRESSURE MICROPHONES

A pressure microphone is a microphone in which the electrical response corresponds to the pressure in a sound wave. The most common pressure microphones are the carbon, condenser, crystal, dynamic, and magnetic microphones.

A. Carbon Microphone

A carbon microphone is a microphone in which operation depends upon the variation in resistance of carbon contacts. The carbon microphone exhibits high sensitivity, which is due to the relay action of the carbon contacts. The carbon microphone is almost universally used in telephonic communications where the prime requisite is sensitivity rather than low distortion and uniform response over the entire audio frequency range.

A sectional view, electrical circuit, mechanical network, and frequency response of a carbon microphone are shown in Fig. 4.4. A displacement of the diaphragm produces a change in the electrical resistance between the

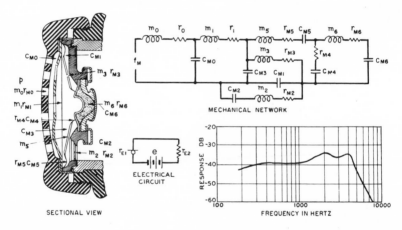

MECHANICAL NETWORK

ELECTRICAL CIRCUIT

SECTIONAL VIEW

FREQUENCY IN HERTZ

Fig. 4.4 Sectional view, the mechanical network of a single-button carbon microphone. In the mechanical network, m_0 and r_{MO} = the mass and mechanical resistance of the holes in the outer grill. C_{MO} = the compliance of the air chamber between the inner and outer grids. m_1 and r_{M1} = the mass and mechanical resistance of the waterproof membrane. C_{M1} = the compliance of the air chamber behind the diaphragm. C_{M3} = compliance of the air chamber behind waterproof membrane and the diaphragm. m_2 and r_{M2} = mass and mechanical resistance of the cloth. m_3 and r_{M3} = mass and mechanical resistance of the hole in the diaphragm. m_5 = mass of the diaphragm. r_{M4} and C_{M4} = mechanical resistance and compliance of the diaphragm to carbon cup couples. C_{M2} = compliance of the air chamber behind the microphone unit. r_{M5} and C_{M5} = the mechanical resistance and compliance of the diaphragm suspension system. m_6, r_{M6}, and C_{M6} = mass, mechanical resistance, and compliance of the carbon cup and granules. f_M = driving force. $f_M = pA$. p = sound pressure. A = the area of the diaphragm. In the electrical circuit: r_{E1} = electrical resistance of the microphone, r_{E2} = electrical resistance of the load. e = polarizing voltage of the battery. The graph shows the voltage developed across the resistor r_{E2} as a function of the frequency for constant sound pressure on the diaphragm.

two terminals of the carbon chamber. The resistance decreases as the motion of the diaphragm compresses the carbon granules and increases as the motion of the diaphragm expands the volume occupied by the carbon granules. Referring to the electrical circuit of Fig. 4.4, the current will be an inverse function of the resistance of the carbon microphone. The voltage developed across the fixed electrical resistor is proportional to the current. Under these conditions the voltage across the fixed electrical resistor is proportional to the current. Under these conditions the voltage across the fixed resistor will correspond to the motion of the diaphragm. Referring to the mechanical network of Fig. 4.4, the vibration of the diaphragm is stiffness-controlled below the resonant frequency. At the resonant frequency, the mechanical resistance provides the controlling element. As a consequence, the amplitude of the diaphragm is proportional to the sound pressure below 4000 hertz. Therefore, the response frequency characteristic is uniform with respect to the frequency in this frequency range as shown in Fig. 4.4. Above the resonant frequency the system becomes mass-controlled and the response falls off quite rapidly.

The microphone shown in Fig. 4.4 is used in the telephone handset of Fig. 7.2.

B. Condenser (Electrostatic) Microphone

A condenser or electrostatic microphone is a microphone in which the operation depends upon the variations in electrical capacitance between two conducting plates. A sectional view, electrical circuit, mechanical network, and frequency response characteristic of a condenser microphone are shown in Fig. 4.5. The condenser microphone consists of a diaphragm separated from a perforated back plate. The diaphragm may be a thin stretched-metal membrane, a plate of a metal, or a metal-coated ceramic or a thin metal-coated plastic membrane. The polarizing voltage is supplied by a source of high voltage. Some plastics exhibit electret properties; that is, the diaphragm is self-polarized and no polarizing voltage is required. An equivalent of up to 200 volts polarization has been obtained in the electret. Voltage and electret have been used for polarizing the diaphragm. See Section 3.5.

Referring to the electrical circuit of Fig. 4.5, the voltage developed across the polarizing resistor is given by:

$$e' = \frac{e_0 r_{E1} C_{E1} \sin \omega t}{C_{E0} \sqrt{(1/\omega C_{E0})^2 + r_{E1}^2}} \qquad (4.8)$$

where e' = voltage developed across the resistor r_{E1}, in statvolts,
   e_0 = polarizing voltage, in statvolts,
   C_{E1} = maximum change in electrical capacitance due to an external applied sinusoidal force, in statfarads,
   $\omega = 2\pi f$,
   f = frequency, in hertz,
   t = time,
   C_{E0} = electrical capacitance of the microphone in repose, in statfarads, and
   r_{E1} = electrical resistance of the polarizing resistor, in statohms.

Equation (4.8) shows that the condenser microphone may be considered as a generator with open-circuit voltage of:

$$e_G = e_0 \left(\frac{C_{E1}}{C_{E0}}\right) \sin \omega t \qquad (4.9)$$

where e_G = open-circuit voltage, in statvolts.

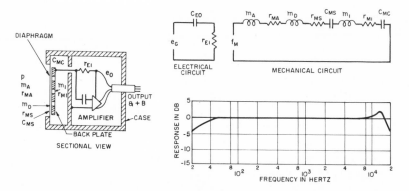

Fig. 4.5 Sectional view, mechanical circuit, electrical circuit, and frequency response characteristic of a pressure condenser microphone. In the mechanical circuit: f_M = driving force. m_A and r_{MA} = mass and mechanical resistance of the air load. m_D = mass of the diaphragm. r_{MS} and C_{MS} = mechanical resistance and compliance of the diaphragm suspension. m_1 and r_{M1} = mass and mechanical resistance of the air in the back plate. C_{MC} = compliance of the air in the case. $f_M = pA$. p = sound pressure on the diaphragm. A = area of the diaphragm. In the electrical circuit: e_G = open circuit voltage. C_{E0} = electrical capacitance. r_{E1} = electrical resistance of the polarizing resistor. e_0 = polarizing voltage. The graph depicts the voltage frequency response characteristic across the resistor r_{E1} for constant sound pressure on the diaphragm.

Furthermore, the electrical impedance of the generator is given by:

$$z_{E0} = \frac{1}{j\omega C_{E0}} \qquad (4.10)$$

where z_{E0} is the electrical impedance of the generator or microphone, in statohms. Equation (4.9) shows that the open-circuit generated voltage is proportional to the amplitude of the diaphragm. Therefore, the system must be stiffness-controlled in order to obtain a uniform frequency response. The frequency amplitude characteristic can be obtained from the mechanical circuit of Fig. 4.5. A typical response frequency characteristic is shown in Fig. 4.5. The high frequency response limit is determined by the fundamental resonant frequency because above this frequency the response falls off rapidly. The response shown in Fig. 4.5 is for constant sound pressure on the diaphragm. In free space an accentuation of pressure and response will occur in the high frequency range as shown for the cylinder of Fig. 1.7. The microphone also becomes directional in the high frequency range as depicted in Fig. 1.7. Equation (4.8) shows that the frequency response falls off when the electrical impedance of the capacitance, C_{E0}, becomes larger than the electrical load resistance, r_{E1}. Therefore, the electrical impedance of the load presented to the condenser microphone must be greater than the internal electrical impedance in order to prevent reduction in the output and frequency discrimination. In this connection the field effect transistor is particularly suitable as an amplifier in Fig. 4.5 because the electrical input impedance is much greater than the internal electrical impedance of the condenser microphone. (See Section 5.2B.) The output electrical impedance of the transistor can be made suitable for transmission over a line by means of a transformer.

The directional characteristic of the pressure condenser microphone is nondirectional as shown in Fig. 4.1A, provided that the diameter of the microphone is less than a quarter of the wavelength. For frequencies above this range the microphone becomes directional .(See Fig. 1.7.)

C. Crystal and Ceramic Microphones

A crystal or ceramic microphone is a microphone in which the operation depends upon the generation of a voltage by the deformation of a crystal or ceramic element having piezoelectric properties. Rochelle salt exhibits the greatest piezoelectric activity of all known crystals. A ceramic element of barium titanate or similar material exhibiting piezoelectric properties may be used, as for example, ammonium dihydrogen phosphate, lead sulphate, lead zirconate, and lead titanate, in place of the Rochelle salt crystal element with some reduction in sensitivity. Otherwise the action and general performance is the same except that a ceramic element will operate at a higher temperature than the Rochelle salt element without suffering permanent damage.

A sectional view, mechanical circuit, electrical circuit, and frequency response characteristic of the crystal microphone are shown in Fig. 4.6. The essential element of the crystal microphone consists of a diaphragm coupled to one corner of the crystal. The other three corners are attached to the case. The motion of the diaphragm deforms the crystal and generates a voltage. The open-circuit voltage developed by the crystal is given by:

$$e_G = Kx \tag{4.11}$$

where e_G = open-circuit voltage, in statvolts,
x = amplitude of the deformation of the crystal, and
K = constant of the crystal.

Equation (4.11) shows that the generated voltage is proportional to the amplitude of the vibration.

The amplitude of the diaphragm is given by:

$$x = \frac{pS}{\omega z_M} \tag{4.12}$$

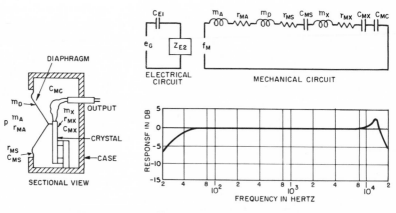

Fig. 4.6 Sectional view, mechanical circuit, electrical circuit and frequency response characteristic of a pressure crystal microphone. In the mechanical circuit: f_M = driving force. m_A and r_{MA} = mass and mechanical resistance of the air load. m_D = mass of the diaphragm. r_{MS} and C_{MS} = mechanical resistance and compliance of the diaphragm suspension. m_X, r_{MX} and C_{MX} = mass, mechanical resistance and compliance of the crystal. C_{MC} = compliance of the air in the case. $f_M = pA$. p = sound pressure on the diaphragm. A = area of the diaphragm. In the electrical circuit: e_G = open circuit generated voltage. C_{E1} = electrical capacitance of the crystal. z_{E2} = electrical impedance of the load. The graph depicts the voltage frequency response characteristic for constant sound pressure on the diaphragm.

where x = amplitude of the diaphragm, in centimeters,
             p = sound pressure, in microbars,
             S = area of the diaphragm, in centimeters,
             ω = $2\pi f$,
             f = frequency, in hertz, and
             z_M = mechanical impedance, in mechanical ohms.

If the mechanical impedance, z_M, is given by:

$$z_M = \frac{1}{2\pi f C_M} \qquad (4.13)$$

where C_M is compliance of the mechanical system, in centimeters per dyne, then the amplitude in Equation (4.12) will be independent of the frequency. Therefore, the vibrating system must be stiffness-controlled in order to obtain a uniform frequency response. The fundamental resonant frequency of the vibrating system of Fig. 4.6 must be placed at the upper limit of the uniform frequency response range. The response frequency characteristic of Fig. 4.6 is for constant sound pressure on the diaphragm. In free space an accentuation of pressure and response will occur in the high frequency range as shown for the cylinder of Fig. 1.7. The microphone also becomes directional in the high frequency range as depicted in Fig. 1.7.

The electrical circuit of the crystal microphone of Fig. 4.6 consists of the open-circuit voltage in series with the capacitance of the crystal and the electrical load.

The capacitance of the element of a crystal or ceramic microphone is of the order of 500 picofarads. Therefore, a relatively long cable of low capacitance may be used between the microphone and the amplifier.

With the advent of the field effect transistor (Section 5.2B) a very small ceramic microphone can be employed by placing the field effect transistor adjacent to the transducing element. Under these conditions the microphone may be used in hearing aids. (See Section 7.3.)

D. Dynamic Microphone

A dynamic microphone is a microphone in which the output results from the motion of a conductor in a magnetic field. A sectional view, mechanical network, electrical circuit, and response frequency characteristic of a dynamic microphone are shown in Fig. 4.7. The dynamic microphone consists of a diaphragm coupled to voice coil located in a magnetic field and a mechanical resistance for controlling the motion of the vibrating system.

The voltage developed in the voice coil is given by:

$$e_G = Bl\dot{x} \qquad (4.14)$$

where e_G = voltage, in abvolts,
   B = flux density in the air gap, in gausses,
   l = length of the voice coil conductor, in centimeters, and
   $\dot{x}$ = velocity of the voice coil, in centimeters per second.

Equation (4.14) shows that in order to provide a uniform frequency response there must be a constant relationship between the velocity of the voice coil and the actuating sound pressure on the diaphragm. The velocity of the diaphragm and voice coil is given by:

$$\dot{x} = \frac{pA}{z_M} \qquad (4.15)$$

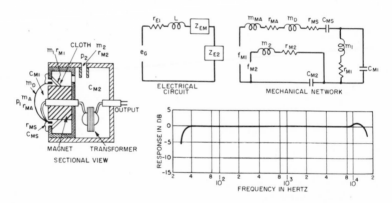

Fig. 4.7 Sectional view, mechanical network, electrical circuit and frequency response characteristic of a pressure dynamic microphone. In the mechanical network: f_{M1} = driving force on the diaphragm. f_{M2} = driving force at the tube. m_A and r_{MA} = mass and mechanical resistance of the air load. m_D = mass of the diaphragm. r_{MS} and C_{MS} = mechanical resistance and compliance of the diaphragm suspension. C_{M1} = compliance of the air behind the diaphragm. m_1 and r_{M1} = mass and mechanical resistance of the cloth damping material. m_2 and r_{M2} = mass and mechanical resistance of the tube. C_{M2} = compliance of the air in the case. $f_{M1} = p_1A$. p_1 = sound pressure on the diaphragm. A = area of the diaphragm. $f_{M2} = p_2A$. p_2 = sound pressure on the tube. In the electrical circuit: e_G = open circuit generated voltage. r_{E1} and L = electrical resistance and inductance of the voice coil. z_{EM} = motional electrical impedance. z_{E2} = load electrical impedance. The graph depicts the open circuit voltage frequency response characteristic for constant sound pressure on the diaphragm.

where $\dot{x}$ = velocity of the diaphragm and voice coil, in centimeters per
          second,
      p = sound pressure, in microbars,
      A = area of the diaphragm, in square centimeters, and
      z_M = total mechanical impedance at f_M of Fig. 4.7.

Equation (4.15) shows that in order to obtain a constant relationship of the
velocity, $\dot{x}$, and the sound pressure, p, the mechanical impedance must be a
mechanical resistance. That is, the vibrating system must be resistance-
controlled. This means that the mechanical resistance r_{M1} must be the controlling factor in the total mechanical impedance.

A typical response frequency characteristic of a dynamic microphone
shown in Fig. 4.7 is for constant sound pressure on the diaphragm. In free
space an accentuation of pressure and response will occur in the high frequency
range as shown for the cylinder of Fig. 1.7. The microphone also becomes
directional in the high frequency range as depicted in Fig. 1.7.

The voice coil of the dynamic microphone is of the order of 10 ohms.
Therefore, as shown in Fig. 4.5, a transformer is used to step up the impedance
to that suitable for transmission over a long line or cable.

The electrical circuit of the dynamic microphone shown in Fig. 4.7 consists
of the open-circuit voltage in series with the electrical impedance of the voice
coil and the electrical load.

E. Magnetic Microphone

A magnetic microphone is a microphone in which the operation depends
upon the variations in the reluctance of a magnetic circuit. A sectional view,
mechanical network, electrical circuit, and response frequency characteristic
of a magnetic microphone are shown in Fig. 4.8. The magnetic microphone
consists of a diaphragm connected to an armature in a polarized magnetic
circuit sometimes referred to as a balanced armature unit. The steady
magnetic flux is supplied by the permanent magnet. Motion of the armature
produces a change in flux through the coil which in turn induces a voltage in
the coil. The voltage developed in the coil is proportional to the velocity of
the armature. Therefore, the conditions for obtaining a uniform response
frequency characteristic are the same as for the dynamic microphone; that is,
there must be a constant relationship between the velocity of the voice coil
and the actuating sound pressure. Referring to the mechanical network, there
are some problems in obtaining a wide frequency range. For example, the
stiffness of the armature must be relatively large in order to maintain centering in the air gap. As a result, the response is attenuated below the fundamental resonant frequency of the system. In the high frequency range

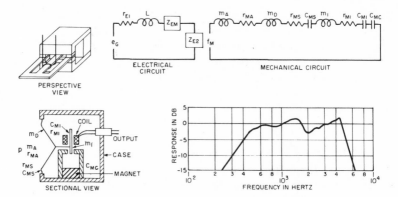

Fig. 4.8 Sectional view, mechanical circuit, electrical circuit and frequency response characteristic of a magnetic microphone. In the mechanical circuit: f_M = driving force. m_A and r_{MA} = mass and mechanical resistance of the air load. m_D = mass of the diaphragm. r_{MS} and C_{MS} = mechanical resistance and compliance of the diaphragm suspension. m_1, r_{M1} and C_{M1} = mass, mechanical resistance and compliance of the armature, including the negative compliance due to the magnetic pull on the armature. C_{MC} = compliance of the air in the case. $f_{M1} = pA$. p = sound pressure on the diaphragm. A = area of the diaphragm. In the electrical circuit: e_G = open circuit generated voltage. r_{E1} and L = electrical resistance and inductance of the coil. z_{EM} = motional electrical impedance. z_{E2} = load electrical impedance. The graph depicts the open circuit voltage frequency response characteristic for constant sound pressure on the diaphragm.

the system becomes mass-controlled and the response falls off in this frequency range. A typical response frequency characteristic is shown in Fig. 4.8. This frequency range is quite satisfactory for the reproduction of speech with a high order of intelligibility. (See Section 17.11.)

The microphone is small in size and light in weight. The sensitivity is relatively high. The magnetic microphone is particularly suitable for speech communication and hearing aid applications. (See Section 7.3.)

The electrical impedance of the coil can be wound with the appropriate size wire to obtain impedances up to several thousand ohms. The electrical circuit of the magnetic microphone shown in Fig. 4.8 consists of the open-circuit voltage of the generator in series with the electrical impedance of the coil and the electrical load.

4.4 VELOCITY MICROPHONE

A velocity microphone is a microphone in which the electrical response corresponds to the particle velocity in a sound wave. The most common velocity microphone is of the ribbon type which will be considered in this section.

Front and sectional views, electrical system, mechanical network, and frequency response characteristic of the velocity microphone are shown in Fig. 4.9. The ribbon is driven by the difference in the sound pressures p_1 and p_2 on the two sides of the ribbon. The magnitudes of the sound pressures on the two sides of the ribbon are the same. Therefore, the difference in sound pressure is due to the difference in phase between the sound pressures on the two sides of the ribbon. The phase difference between the two sides of the ribbon is due to the difference in the sound path between the sides of the ribbon. The difference in the sound path between the two sides of the ribbon is depicted in Fig. 4.10 for sound incident at $0°$, $45°$, and $90°$. The difference in the sound path, which is due to the direction of the impinging sound wave,

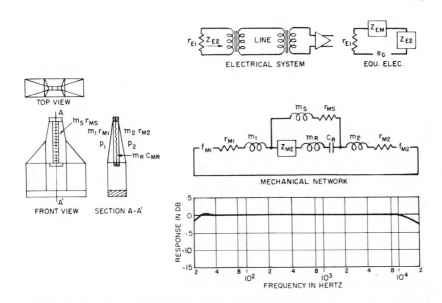

Fig. 4.9 Front, top and sectional views, mechanical network, electrical system, equivalent electrical circuit and the frequency response characteristic of a velocity microphone. In the mechanical network: f_{M1} = driving force on the front of the ribbon. f_{M2} = driving force on the back of the ribbon. m_1 and r_{M1} = mass and mechanical resistance of the air load on the front of the ribbon. m_2 and r_{M2} = mass and mechanical resistance of the air load on the back of the ribbon. m_R and C_{MR} = mass and compliance of the ribbon. m_S and r_{MS} = mass and mechanical resistance of the air in the slit between the ribbon and the pole pieces. z_{ME} = mechanical impedance due to the electrical system. $f_{M1} = p_1 A$. p_1 = sound pressure on the front of the ribbon. A = area of the ribbon. $f_{M2} = p_2 A$. p_2 = sound pressure on the back of the ribbon. In the electrical circuit: e_G = open circuit generated voltage. z_{EM} = motional electrical impedance. z_{E2} = load electrical impedance. r_{E1} = generator electrical resistance. The graph depicts the open circuit voltage frequency response characteristic for constant sound pressure in free space.

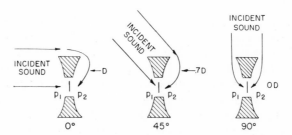

Fig. 4.10 The difference in sound path between the front and back of the ribbon for sound incident at 0°, 45° and 90°. p_1 and p_2 = sound pressures on the front and back of the ribbon.

is a cosine function. The difference in pressure between the two sides of the ribbon is given by:

$$\Delta p = 2p_m \, (\cos \omega t) \left(\sin \frac{kD \cos \theta}{2} \right) \qquad (4.16)$$

where Δp = difference in sound pressure between the two sides of the ribbon, in microbars,

p_m = amplitude of the sound pressure, in microbars,

ω = $2\pi f$,

f = frequency, in hertz,

t = time, in seconds,

k = $2\pi/\lambda$,

λ = wavelength, in centimeters,

D = sound path between the two sides of the ribbon, in centimeters, and

θ = angle between the direction of the incident sound and the normal to the plane of the ribbon.

The difference in sound pressure between the two sides of the ribbon as a function of the acoustic path between the front and back of the ribbon is shown in Fig. 4.11.

The velocity of the ribbon is given by:

$$\dot{x} = \frac{\Delta p A}{z_M} \qquad (4.17)$$

where $\dot{x}$ = velocity of the ribbon, in centimeters per second,

A = area of the ribbon, in square centimeters, and

z_M = total mechanical impedance, in mechanical ohms.

Above the mechanical resonance of the ribbon the controlling mechanical impedance is due to the mass of the air load and the mass of the ribbon,

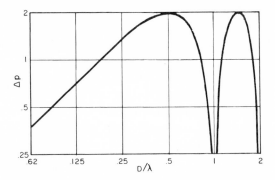

Fig. 4.11 The difference in sound pressure Δp between the front and back of the ribbon as a function of D/λ. D = sound path from the front to the back of the ribbon. λ = wavelength.

because the mechanical impedance due to the electrical load is small and may be neglected, and the mechanical impedance due to the slit between the ribbon and the pole pieces is large and may be neglected. Under these conditions Equation (4.17) becomes:

$$\dot{x} = \frac{\Delta p A}{j\omega(m_R + m_A)} \tag{4.18}$$

where
m_R = mass of the ribbon, and
m_A = mass of the air load.

Substitution of the value of Δp from Equation (4.16) into Equation (4.18) gives a result of:

$$\dot{x} = \frac{2p_m A \, (\cos \omega t) \, \{\sin \, [(kD \cos \theta)/2]\}}{j\omega(m_R + m_A)} \tag{4.19}$$

The frequency velocity response characteristic of the ribbon obtained from Equation (4.19) for $0 = 0$ is shown in Fig. 4.12. An examination of Fig. 4.12 shows that uniform response is obtained for small values of the ratio D/λ.

The voltage developed by the motion of the ribbon is given by the equation:

$$e_G = Bl\dot{x} \tag{4.20}$$

where
e_G = generated voltage, in abvolts,
B = flux density in the air gap, in gausses,
l = length of the ribbon, in centimeters, and
$\dot{x}$ = velocity of the ribbon, in centimeters per second, as given by Equation (4.19).

The frequency voltage response is the same as the frequency velocity response of the ribbon of Fig. 4.12 save for the constant Bl. The response

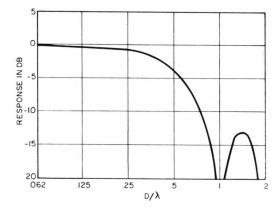

Fig. 4.12 The velocity frequency characteristic of the ribbon as function of D/λ. D = distance from the front to the back of the ribbon. λ = wavelength.

frequency characteristic of a velocity microphone of the type shown in Fig. 4.9 is shown in Fig. 4.12.

The equivalent electrical circuit of the velocity microphone consists of the open-circuit voltage of the generator in series with the electrical resistance and motional electrical impedance of the ribbon and the electrical impedance of the electrical load.

The directional pattern of the velocity microphone is another important consideration in the performance. The geometry for the directional characteristics in the horizontal and vertical planes is illustrated in Fig. 4.13. The directional characteristic in the horizontal plane as depicted in Fig. 4.13

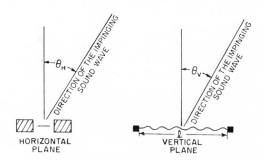

Fig. 4.13 The geometry for directional characteristics of a ribbon velocity microphone. θ_H = angle between the direction of the incident sound and the normal to the ribbon. θ_V = angle between the direction of the incident sound and the normal to the ribbon.

is given by:

$$R_H = \frac{\sin\,[(kD/2)\cos\theta_H]}{kD/2} \tag{4.21}$$

where R_H is the ratio of the response for angle θ_H to the response for $\theta_H = 0$, and θ_H is the angle between the direction of the impinging sound wave in the horizontal plane and the normal to the plane of the ribbon. For small values of kD, Equation (4.21) becomes:

$$R_H = \cos\theta_H \tag{4.22}$$

The directional characteristic in the vertical plane as depicted in Fig. 4.13 is given by:

$$R_V = \frac{\sin\,[(\pi l/\lambda)\sin\theta_V]}{(\pi l/\lambda)\sin\theta_V} \cdot \frac{\sin\,[(kD/2)\cos\theta_V]}{kD/2} \tag{4.23}$$

where R_V is the ratio of the response for an angle θ_V to the response for an angle $\theta_V = 0$. For small values of ratio l/λ and the product kD, Equation (4.23) becomes:

$$R_V = \cos\theta_V \tag{4.24}$$

4.5 UNIDIRECTIONAL MICROPHONES

A unidirectional microphone is a microphone with a substantially uni-directional pattern over the response frequency range. The first unidirectional microphones were combinations of pressure and velocity microphones. Later single-element unidirectional microphones were developed. Both types of unidirectional microphone will be described in this section.

A. Combination Unidirectional Microphone

The combination unidirectional microphone consists of a bidirectional velocity microphone and a pressure microphone. The voltage output of the combination unidirectional microphone is given by:

$$e_U = e_P + e_V \cos\theta \tag{4.25}$$

where e_U = voltage output of the unidirectional microphone,
e_P = voltage output of the pressure element,
e_V = voltage output of the velocity element for $\theta = 0$, and
θ = angle between the direction of the incident sound and the major axis of the microphone.

If $e_P = e_V$, then the directional characteristic is the cardioid pattern of Fig. 4.1C. If $e_P = 0.3e_V$, then the directional characteristic is the limacon pattern of Fig. 4.1D.

B. Single-Element Ribbon Unidirectional Microphone

A single-element ribbon unidirectional microphone consists of a ribbon pressure gradient microphone with an appropriate acoustical delay system. Perspective views of a single-element ribbon-type unidirectional microphone are shown in Fig. 4.14. The ribbon is connected to a folded damped pipe. A front view, sectional view, electrical system, mechanical network, equivalent electrical circuit, and frequency response characteristic of a single-element unidirectional microphone are shown in Fig. 4.15. The difference in pressure between the two sides of the ribbon is due to difference in phase between the two pressures p_1 and p_2 as well as the additional phase shift due to the mechanical network r_{M2}, m_2, and r_{M3}. The mechanical network introduces a difference in phase between p_2 and p_2' of Figs. 4.15 and 4.16. The difference in the acoustic path between the two sides of the ribbon for 0°, 90°, and 180° is depicted in Fig. 4.16. The difference in the acoustic path produced by the aperture m_2 and r_{M2} and the damped pipe r_{M3} is the same for all angles of the incident sound. The outside acoustic path is a function of the angle of the incident sound. The total difference in the acoustic path between the two sides of the ribbon may be expressed as:

$$D_T = \frac{D}{2} + \frac{D}{2} \cos \theta \qquad (4.26)$$

where D_T = total difference in acoustic path between the two sides of the ribbon.

The difference in pressure between the two sides of the ribbon for the

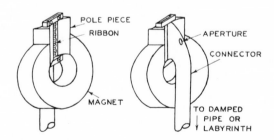

Fig. 4.14   The elements of a single element ribbon unidirectional microphone.

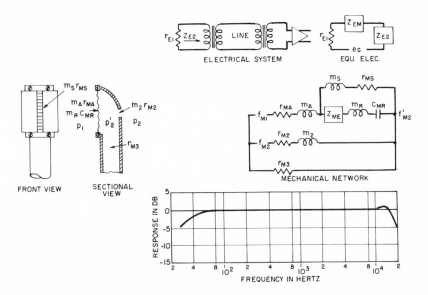

Fig. 4.15 Front and sectional views, mechanical network, electrical system, equivalent electrical circuit and frequency response characteristic of a single-element ribbon unidirectional microphone. In the mechanical network: f_{M1} = driving force on the front of the ribbon. f_{M2} = driving force on the aperture. m_A and r_{MA} = mass and mechanical resistance of the air load on the front of the ribbon. m_R and C_{MR} = mass and compliance of the ribbon. m_S and r_{MS} = mass and mechanical resistance of the air in the slit between the ribbon and the pole pieces. z_{ME} = mechanical impedance due to the electrical system. m_2 and r_{M2} = mass and mechanical resistance of the aperture in the back cover. r_{M3} = mechanical resistance of the damped pipe. $f_{M1} = p_1 A$. p_1 = sound pressure on the front of the ribbon. A = area of the ribbon. $f_{M2} = p_2 A$. p_2 = sound pressure on the aperture. $f'_{M2} = p'_2 A$. p'_2 = sound pressure inside the pipe. In the electrical circuit: e_G = open circuit generated voltage. r_{E1} = generator electrical resistance. z_{EM} = motional electrical impedance. z_{E2} = load electrical impedance. The graph depicts the open circuit voltage frequency response characteristic for constant sound pressure in free space.

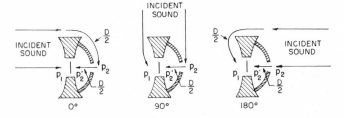

Fig. 4.16 The difference in the sound path between the front and back of the single element ribbon unidirectional microphone for sound incident at 0°, 90° and 180°. p_1 = sound pressure on the front of the ribbon. p_2 = sound pressure on the aperture. p'_2 = sound pressure on the back of the ribbon.

cardioid directional characteristic is given by:

$$\Delta p = 2p_m \left[\cos \omega t\right] \sin \left[kD \frac{1 + \cos \theta}{4}\right] \tag{4.27}$$

where Δp = difference in pressure between the two sides of the ribbon, in microbars,

p_m = amplitude of the sound pressure, in microbars,

$\omega = 2\pi f$,

f = frequency, in hertz,

t = time, in seconds,

$k = 2\pi/\lambda$,

λ = wavelength, in centimeters,

D = acoustic path, in centimeters, and

θ = angle between the direction of the incident sound and the axis of the microphone.

The difference in pressure between the two sides of the ribbon is depicted in Fig. 4.16.

The velocity of the ribbon is given by:

$$\dot{x} = \frac{\Delta p A}{z_M} \tag{4.28}$$

where $\dot{x}$ = velocity of the ribbon, in centimeters per second,

A = area of the ribbon in square centimeters, and

z_M = mechanical impedance, in mechanical ohms.

Above the mechanical resonance of the ribbon the controlling mechanical impedance is due to the mass of the air load and the mass of the ribbon, because the mechanical impedance due to the electrical load is small and may be neglected, and the mechanical impedance due to the leakage in the slit between the ribbon and the pole pieces is large and may be neglected. Under these conditions the velocity of the ribbon is given by:

$$\dot{x} = \frac{\Delta p A}{j\omega(m_R + m_A)} \tag{4.29}$$

where m_R is mass of the ribbon, and m_A is mass of the air load.

Substitution of Δp from Equation (4.27) into Equation (4.29) gives the velocity of the ribbon as:

$$\dot{x} = \frac{2p_m A \left[\cos \omega t\right] \sin \left[kD(1 + \cos \theta)/4\right]}{j\omega(m_R + m_A)} \tag{4.30}$$

The frequency velocity response characteristic of the ribbon according to Equation (4.30) is shown in Fig. 4.12. An examination shows that a uniform response is obtained for small values of the ratio D/λ. The voltage generated by the ribbon is given by Equation (4.20). (A typical frequency response characteristic of the single-element unidirectional microphone is shown in Fig. 4.15.)

The equivalent electrical circuit of the single-element unidirectional microphone consists of the open-circuit voltage of the generator in series with the electrical resistance and motional electrical impedance of the ribbon and the electrical impedance of the electrical load.

Another form of the single-element ribbon unidirectional microphone in which the maximum response is along the cylindrical axis of the microphone is shown in Fig. 4.17. The mechanical network of the vibrating system is the

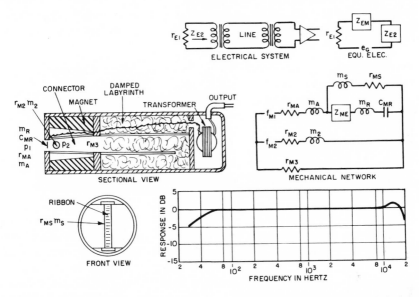

Fig. 4.17 Front and sectional views, mechanical network, electrical system, equivalent electrical circuit and frequency response characteristic of a single element ribbon unidirectional microphone of the axial type. In the mechanical network: f_{M1} = driving force on the front of the ribbon. f_{M2} = driving force on the aperture. m_A and r_{MA} = mass and mechanical resistance of the air load on the front of the ribbon. m_R and C_{MR} = mass and compliance of the ribbon. m_S and r_{MS} = mass and mechanical resistance of the air in the slit between the ribbon and the pole pieces. z_{ME} = mechanical impedance due to the electrical system. m_2 and r_{M2} = mass and mechanical resistance of the aperture. r_{M3} = mechanical resistance of the labyrinth. $f_{M1} = p_1 A$. p_1 = sound pressure on the front of the ribbon. A = area of the ribbon. $f_{M2} = p_2 A$. p_2 = sound pressure on the aperture. In the electrical circuit: e_G = open circuit generated voltage. r_{E1} = generator electrical resistance. z_{EM} = motional electrical impedance. z_{E2} = load electrical impedance. The graph depicts the open circuit voltage frequency response characteristic for constant sound pressure in free space.

same as for the microphone of Fig. 4.15. A typical frequency response characteristic response is shown in Fig. 4.17. Any unidirectional characteristic between the cardioid and limacon of highest directivity of Fig. 4.1 may be obtained by suitable selection of the parameters of the microphone. The constants of the microphone are usually selected so that the directional characteristic is between the cardioid of Fig. 4.1C and the limacon of maximum directivity of Fig. 4.1D.

C. Single-Element Dynamic Unidirectional Microphone

A sectional view, mechanical network, electrical circuit, and frequency response characteristic of a single-element dynamic unidirectional microphone are shown in Fig. 4.18. The performance is similar to that of the single-element ribbon unidirectional microphone of Section 4.5B.

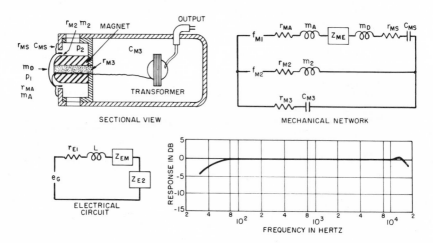

Fig. 4.18 Front and sectional views, mechanical network, electrical circuit and response frequency characteristic of a single element dynamic unidirectional microphone. In the mechanical circuit: f_{M1} = driving force on the diaphragm. f_{M2} = driving force at the voice coil aperture. m_A and r_{MA} = mass and mechanical resistance of the air load on the diaphragm. m_D = mass of the diaphragm. r_{MS} and C_{MS} = mechanical resistance and compliance of the suspension system. m_2 and r_{M2} = mass and mechanical resistance of the voice coil aperture. z_{ME} = mechanical impedance due to the electrical system. r_{M3} = mechanical resistance in the center pole. C_{M3} = compliance of the air in the case. $f_{M1} = p_1 A$. p_1 = sound pressure on the diaphragm. A = area of the diaphragm. $f_{M2} = p_2 A$. p_2 = sound pressure at the voice coil aperture. In the electrical circuit: e_G = open circuit generated voltage. r_{E1} and L = electrical resistance and inductance of the voice coil. z_{EM} = motional electrical impedance. z_{E2} = load electrical impedance. The graph depicts the open-circuit voltage frequency response characteristic for constant sound pressure in free space.

A typical frequency response characteristic is shown in Fig. 4.18. Any unidirectional directivity pattern may be obtained between the cardioid and limacon of highest directivity of Fig. 4.1 by suitable selection of the parameters of the microphone.

The electrical circuit of the unidirectional dynamic microphone of Fig. 4.18 is the same as that for the pressure microphone of Fig. 4.7.

D. Single-Element Condenser (Electrostatic) Unidirectional Microphone

A sectional view, mechanical network, electrical circuit, and frequency response of a single element unidirectional condenser or electrostatic microphone are shown in Fig. 4.19. As in the case of the unidirectional microphones of Figs. 4.15, 4.17, and 4.18, the actuating pressure is proportional to the frequency. In the case of the electrostatic system the developed voltage is proportional to the amplitude of the diaphragm. (See Section 4.3B.) Under

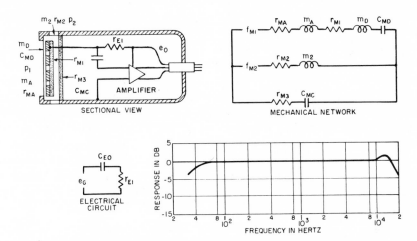

Fig. 4.19 Sectional view, mechanical network, electrical circuit and frequency response characteristic of a single element condenser unidirectional microphone. In the mechanical network: f_{M1} = driving force on the diaphragm. f_{M2} = driving force on the aperture. m_A and r_{MA} = mass and mechanical resistance of the air load on the diaphragm. m_D and C_D = mass and compliance of the diaphragm. r_{M1} = mechanical resistance control of the diaphragm. m_2 and r_{M2} = mass and mechanical resistance of the aperture. r_{M3} = mechanical resistance. C_{MC} = compliance of the air in the case. $f_{M1} = p_1 A$. p_1 = sound pressure on the diaphragm. A = area of the diaphragm. $f_{M2} = p_2 A$. p_2 = sound pressure at the aperture. In the electrical circuit: e_G = open circuit generated voltage. C_{EO} = electrical capacitance of the condenser. r_{E1} = electrical resistance of the load. The graph depicts the open circuit voltage frequency response characteristic for constant sound pressure in free space.

these conditions the diaphragm system must be resistance-controlled in order to obtain a uniform response with respect to the frequency. This control is accomplished by means of the mechanical resistance r_{M1} of Fig. 4.19. Aside from this departure, the action is the same as that of the microphone of Fig. 4.18. A typical frequency response characteristic is shown in Fig. 4.19. Any unidirectional directivity pattern may be obtained between the cardioid and limacon of highest directivity of Fig. 4.1 by suitable selection of the parameters of the microphone.

The electrical circuit of the unidirectional condenser microphone of Fig. 4.19 is the same as that for the pressure electrostatic microphone of Section 4.3B. The polarization systems are the same as for the pressure electrostatic microphone.

A sectional view and the electrical systems of a single unit condenser or electrostatic microphone which may be operated as a nondirectional, bidirectional, unidirectional, or polydirectional microphone are shown in Fig. 4.20. The microphone consists of two thin metal-coated plastic diaphragms and center electrode. In the pressure mode of operation of Fig. 4.20A, the motion of the two diaphragms toward and away from the central electrode is used. For this operation the electrical connection of Fig. 4.20A must be used. The free resonance of the diaphragms occurs at a very low frequency. Therefore, in the pressure operation the space behind the diaphragms is the controlling impedance. Since this is a compliance, the system is compliance-controlled. Therefore, the voltage output is proportional to the sound pressure and independent of the frequency. The directional pattern is that of Fig. 4.1A. The resonance usually occurs at the upper part of the audio frequency range. The amplitude is damped by the fine damping tubes in the center electrode. In the velocity operation of Fig. 4.20B, the motions of the two diaphragms are in unison in the same direction. The diaphragms are driven by the difference in sound pressure between the two sides of the

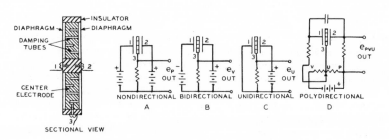

Fig. 4.20 Sectional views and electrical systems of a single unit condenser microphone. A. Nondirectional operation. B. Bidirectional cosine operation. C. Unidirectional cardioid operation. D. Polydirectional operation.

microphone. (See Section 4.4.) For this operation the polarity of the polarizing batteries is reversed from that of the pressure operation. The system is controlled by the mechanical resistance of the small tubes in the center electrode connecting the backs of the two diaphragms. The actuating pressure is proportional to the frequency. Therefore, for constant sound pressure in free space, the amplitude will be independent of the frequency. The voltage output will also be independent of the frequency. The directional pattern is that of Fig. 4.1B. Zero output occurs for sound arriving in the plane of the diaphragm. If the two microphones of Figs. 4.20A and B are combined, the directional pattern will be that of the cardioid of Fig. 4.1C. (See Section 4.5A.) The electrical system for the cardioid directional pattern is shown in Fig. 4.20C. The maximum output occurs for sound arriving normal to diaphragm 2 because the pressure and velocity components are additive. The output is very low for sound arriving normal to diaphragm 1 because the pressure and velocity components cancel each other. For sound arriving in the plane of the diaphragms, there is no velocity component and the output is down 6 dB from the maximum output. The electrical system for a polydirectional microphone is shown in Fig. 4.20. By moving the contactor from the V to U to P, one may obtain all the directional patterns from bidirectional to unidirectional to nondirectional.

The amplifier for the microphone of Fig. 4.20 is the same as that described in Section 4.3B.

The frequency response is essentially the same as that shown in Fig. 4.19. There may be some accentuation in high frequency response in the case of reduced damping and diffraction incurred by a large-diameter element. (See Section 1.3T.)

4.6 HIGHER ORDER GRADIENT MICROPHONES

First order pressure gradient microphones of the velocity and unidirectional types have been described in the preceding sections. The responses of these microphones correspond to the first order of the pressure gradient in a sound wave. The responses of higher order gradient microphones correspond to the order of the pressure gradient in the sound wave.

A. Second Order Gradient Microphones

A first order gradient bidirectional microphone of the ribbon type is depicted in Fig. 4.21A. When two of these microphones are connected in opposition as shown in Fig. 4.21B, the result is a second order pressure gradient

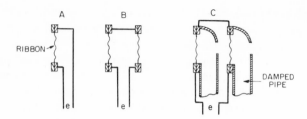

Fig. 4.21   Gradient microphones of the ribbon type. A. Bidirectional first order gradient microphone. B. Bidirectional second order gradient microphone. C. Unidirectional second order gradient microphone. e = open circuit voltage output.

bidirectional microphone. The directional characteristic is given by:

$$R_D = \cos^2 \theta \qquad (4.31)$$

where R_D = response for the angle θ, and
$\quad\quad\quad \theta$ = angle between the direction of the impinging sound wave and the normal to the planes of the ribbons.

The directional characteristic is shown in Fig. 4.2C.

Second order gradient microphones may also be constructed using other types of transducers with both sides of the diaphragms exposed to the sound wave and the outputs connected in opposition.

A second order gradient unidirectional microphone consisting of two cardioid microphones of the type shown in Figs. 4.15 or 4.17 connected in opposition is shown in Fig. 4.21C. The directional characteristic is given by:

$$R_D = (1 + \cos \theta) \cos \theta \qquad (4.32)$$

where R_D = response at the angle θ, and
$\quad\quad\quad \theta$ = angle between the direction of the impinging sound wave and the normal to the planes of the ribbons.

The directional characteristic is shown in Fig. 4.3B.

A second order unidirectional microphone has been commercialized employing two unidirectional microphones of the type shown in Fig. 4.17 and an appropriate electrical network.

B. Noise Discrimination of Gradient Microphones

Gradient microphones of the order one and higher are directional. Therefore, these microphones discriminate against sounds from random directions.

From Equation (4.6), the directional efficiency of a gradient microphone is given by:

$$D.E. = \frac{1}{2n + 1} \tag{4.33}$$

where $D.E.$ = directional efficiency, and
                  n = order of the gradient.

A further increase in discrimination against noise and other undesired sounds may be obtained if a gradient microphone is used as a close-talking microphone. The response of gradient microphones of order zero, one, and two to a small source will be developed in the exposition which follows. The parameters and variables will be selected so that the response of all three microphones is uniform with respect to frequency when the distance between the sound source and the microphones is large.

The response of a zero order gradient microphone is given by:

$$e_0 = \frac{S_0}{r} \tag{4.34}$$

where e_0 = output of the microphone, in volts,
        S_0 = constant involving the strength of the sound source and the sensitivity of the microphone, and
        r = distance between the sound source and the microphone.

The axial response of a first order gradient microphone is given by:

$$e_1 = S_1 \left(\frac{1}{r} + \frac{j}{kr^2} \right) \tag{4.35}$$

where e_1 = output of the microphone, in volts,
        S_1 = constant involving the strength of the sound source and the sensitivity of the microphone,
        $k = 2\pi/\lambda$,
        λ = wavelength, in centimeters, and
        j = the operator $\sqrt{-1}$.

The axial response of a second order gradient microphone is given by:

$$e_2 = S_2 \left(\frac{1}{r} + \frac{1}{2k^2 r^3} + \frac{j}{kr^2} \right) \tag{4.36}$$

where e_2 = output of the microphone, in volts, and
        S_2 = constant involving the strength of the sound source and the sensitivity of the microphone.

Equation (4.34) shows that the response of the zero order microphone is independent of the frequency regardless of the distance. Equations (4.35) and (4.36) show that the low frequency response of the first and second order gradient microphones will be accentuated when the ratio λ/r is greater than unity and increases as the ratio increases. Furthermore, the accentuation in low frequency response for increase of the ratio λ/r is greater for the second order gradient microphone. The accentuation in response feature of a gradient microphone may be used to obtain high discrimination against unwanted sounds if the microphone is used as a close-talking microphone and the unwanted sounds originate at a distance from the microphone. For example, if the distance between the mouth and the microphone is $\frac{3}{4}$ inch, which is the average distance for a close-talking microphone, the frequency response characteristic of zero, first, and second order gradient microphones will be as shown in Fig. 4.22A. The response of the gradient microphones is accentuated at the low frequencies. If compensation is introduced so that the response of all three becomes uniform with respect to frequency for the $\frac{3}{4}$-inch distance from the small sound source, the response frequency characteristics for distant sounds will be as shown in Fig. 4.21B. These characteristics show the discrimination against distant axial sounds by the first and second order gradient microphones as compared to a pressure or zero order gradient microphone. These characteristics apply to all first and second order gradient microphones.

For the sake of simplicity ribbon type gradient microphones have been depicted in Fig. 4.22. As this chapter has indicated, any type of microphone may be used to obtain zero and first order gradient microphones. This may be extended to higher order gradient microphones.

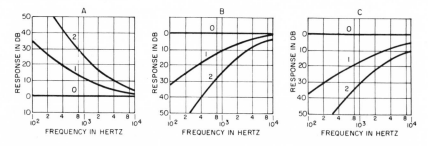

Fig. 4.22 A. Response frequency characteristics of zero, first and second order gradient microphones for a distance from the sound source of $\frac{3}{4}$ inch. B. Response frequency characteristics of zero, first and second order microphones to a plane wave when the microphones are compensated to provide uniform frequency response characteristics at $\frac{3}{4}$ inch. C. Compensation same as B. The frequency response characteristics depict the response of the three microphones to random sounds originating in random directions.

In general, noise and unwanted sounds originate in random directions. Under these conditions additional discrimination will be introduced by the directional pattern. (See Section 4.2.) The response of zero, first, and second order gradient microphones, compensated for uniform response at $\frac{3}{4}$-inch distance, to distant sounds originating in random directions is shown in Fig. 4.22C.

4.7 WAVE-TYPE MICROPHONES

Directional microphones may be divided into two classes, as follows: first, wave-type microphones which depend for directivity upon wave interference, and second, gradient-type microphones which depend for directivity upon the difference in pressure or powers of the difference in pressure between two points. In the first class of microphone, in which the directivity depends in some way upon wave interference, if any semblance of directivity is obtained, the dimensions of the microphone must be comparable to the wavelength of the sound wave. The dimensions of gradient microphones as contrasted to wave-type microphones are small compared to the wavelength. Two examples of wave-type microphones will be considered in the section which follows.

A. Parabolic Reflector Microphone

Reflectors have been used for years for concentrating and amplifying all types of wave propagation. The surface of the parabolic reflector is shaped so that the various pencils of incident sound parallel to the axis are reflected to one point, called the focus, as shown in Fig. 4.23. To obtain an appreciable

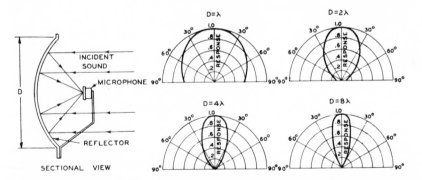

Fig. 4.23 Sectional view and directional characteristics of a parabolic reflector microphone with the microphone slightly out of focus as a function of the ratio of the diameter of the reflector to the wavelength. λ = wavelength. D = diameter of the reflector.

gain in pressure at the focus, the reflector must be large compared to the wavelength of the incident sound. This requirement of size must also be satisfied in order to obtain sharp directional characteristics. If this condition is satisfied at the low frequencies, the size of the reflector becomes prohibitive to be used with facility.

A cross-sectional view of a parabolic reflector and a pressure microphone located at the focus is shown in Fig. 4.23. When the microphone is located at the focus, the gain at the high frequencies is considerably greater than at the mid-frequency range. The accentuation in high frequency response may be overcome by moving the microphone slightly out of focus. This expedient also tends to broaden the sharp directional characteristics at the high frequencies.

The directional characteristics of a parabolic reflector microphone are shown in Fig. 4.23. These characteristics show that the diameter of the reflector must be greater than the wavelength to obtain any useful directivity.

B. Line Microphone

A line microphone consists of a series of sound-pickup points positioned along a line with the outputs connected to an acoustoelectric transducer. One form of the line microphone consists of a series of small pipes with the open ends as sound-pickup points equally spaced along a line and the other ends connected to an acoustoelectric transducer as shown in Fig. 4.24.

The directional characteristic is given by:

$$R_D = \frac{\sin\left[(\pi/\lambda)(l - l\cos\theta)\right]}{(\pi/\lambda)(l - l\cos\theta)} \tag{4.37}$$

where R_D = response for the angle θ,
  l = length of the line, in centimeters,
  λ = wavelength, in centimeters, and
  θ = angle between the direction of the impinging sound wave and the axis of the line.

The directional characteristics of the microphone of Fig. 4.24 for various ratios of the length of the line to the wavelength are shown in Fig. 4.25.

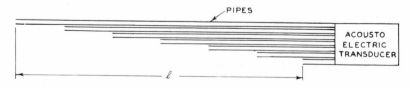

Fig. 4.24 The elements of a line microphone consisting of pipes connected to an acoustoelectric transducer. *l* = length of the line.

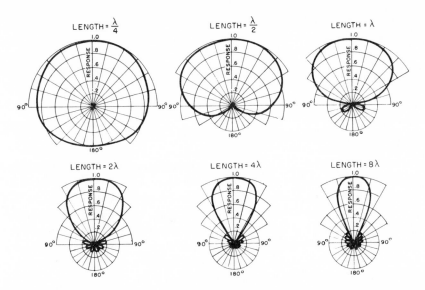

Fig. 4.25   The directional characteristics of the line microphone shown in Fig. 4.24 as a function of the ratio of the length of the line to the wavelength. $\theta = 0$ corresponds to the axis of the line. λ = wavelength. l = length of the line.

The directional characteristics are surfaces of revolution about the line as an axis. The line microphone is useful for collecting sounds arriving at directions making small angles with the microphone axis.

Various transducers, such as ribbon, dynamic, and condenser, have been used with line microphones.

4.8 SPECIAL-APPLICATION MICROPHONES

Lavalier, boom, probe-tube, and ratio microphones are among the special-application microphones that will be described in this section.

A. Lavalier, Boom, and Probe-Tube Microphones

A lavalier microphone is a term used to designate a small microphone supported by means of a slim band around the neck in the form of a pendant. A boom-type microphone is usually used in conjunction with earphones and a headband in which the boom supporting a small microphone in front of the mouth is attached to the headband. A probe-tube microphone consists of a slender tube connected to a small microphone element with the open end positioned in front of the mouth. The microphone element may be supported by the earphone headband or if earphones are not worn by a separate headband.

The principal virtue of personal microphones as exemplified by the lavalier, probe-tube, and boom microphones is to allow a person to move about freely without introducing any change in output from the microphone as would occur if a stationary microphone were used.

B. Radio Microphone

A radio microphone consists of a microphone, small radio transmitter, antenna, and battery assembly, and a radio receiver as depicted in Fig. 4.26. The use of solid state elements makes it possible to house the elements of the transmitter and the battery in a small case. The antenna is in the form of a wire 6 to 12 inches in length. The microphone-transmitter is usually worn as a lavalier microphone. A very wide dynamic range automatic gain control system is used in the receiver to compensate for variations in transmission due to standing wave systems.

4.9 MICROPHONE PERFORMANCE REQUIREMENTS FOR HIGH FIDELITY SOUND REPRODUCTION

The term high fidelity sound reproduction is used to designate a superior order of performance. The characteristics involved in describing the performance of a microphone are frequency response, directivity, nonlinear distortion, transient response, electrical impedance, sensitivity, and noise.

A. Frequency Response

The frequency response of a microphone is the open-circuit electrical voltage output, in volts, for constant sound pressure in an undisturbed sound field for sound arriving on the axis of the microphone, as a function of the frequency. For microphones employing dynamic, ribbon, crystal, and electrostatic transducers there is no problem in achieving uniform response over the

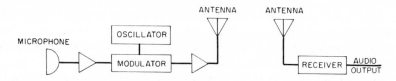

Fig. 4.26   Elements of a radio microphone.

entire audio frequency range, as has been shown in this chapter. However, as has been suggested here, there are some applications where a limited frequency response is indicated.

B. Directivity

The polar directional characteristic of a microphone is the response as a function of the angle of the impinging sound wave with respect to some axis of the microphone. There is no problem in achieving a uniform directional characteristic in a high-quality microphone.

C. Nonlinear Distortion

The frequency nonlinear distortion characteristic of a microphone is the total nonlinear distortion, for actuation by a sine pressure sound wave, as a function of the frequency. The nonlinear distortion generated in a microphone is negligible for sound levels encountered in the audio amplitude range.

D. Transient Response

The transient response of a microphone refers to the faithfulness of response to a sudden change in the sound input. For high-quality microphones with uniform response over the entire audio frequency range, the transient response is excellent.

E. Electrical Impedance

The frequency electrical impedance characteristic of a microphone is the electrical impedance as a function of the frequency. The EIA (Electrical Industry Association) standard impedance is 150 ohms. Other common electrical impedances are 30 and 250 ohms.

F. Sensitivity

The sensitivity of a microphone is open-circuit voltage output, in volts, for a free-field sound pressure of one microbar. The EIA Standard defines the Effective Output Level of a microphone as the open-circuit voltage output at the nominal output impedance, in decibels (dBm) relative to one volt, for a sound pressure of 10 microbars.

The EIA Standard defines the system rating G_M of a microphone as the ratio in decibels relative to 0.001 watt per microbar of the maximum electrical power from the microphone to the square of the undisturbed sound field pressure in a plane progressive sound wave at the microphone position. The system rating of a microphone is expressed as:

$$G_M = 20 \log_{10} \frac{e}{p} - 10 \log_{10} r_E - 50 \qquad (4.38)$$

where G_M = system rating of the microphone, in decibels,
$\quad\quad\quad e$ = open-circuit voltage output of the microphone, in volts,
$\quad\quad\quad p$ = free-field sound pressure, in microbars, and
$\quad\quad\quad r_E$ = electrical impedance of the microphone, in ohms.

G. Noise

There are two principal sources of noise in a microphone, namely, the thermal agitation noise and hum pickup from ac fields. The thermal agitation noise generated in the conductors or the polarizing and biasing resistors can be calculated from Equation 5.6 in Section 5.9. For most high-quality microphones the thermal agitation noise for the audio frequency range is of the order of an equivalent sound pressure of 10 dB. This is about 10 dB below the ambient noise in a very quite studio. (See Section 15.12.) Therefore, for most practical applications the thermal agitation noise generated in a microphone is not a factor in the pickup of sound.

For determining the hum level of a microphone, an arbitrary standard 60-hertz ac field of 10^{-3} gauss has been established as a reference. The hum level is calculated in the same fashion as the Effective Output Level (dBm) using as the output voltage, the voltage produced by the standard field. For a well-designed microphone the hum pickup is not a factor in the pickup of sound.

H. Wind Noise and Wind Screens

Microphones used outdoors are often excited by wind noise. Wind noise is an impulse-type of excitation. The amplitudes of the components are inversely proportional to the frequency. Therefore, attentuation of the low frequency response reduces the wind noise response. Screens in the form of cloth and porous plastics will reduce the wind excitation. The effectiveness of the screen is a direct function of its size and acoustical resistance, with the acoustical resistance tending to reduce the high frequency response. Nevertheless, the trend seems to be to use relatively small porous plastic wind screens with as high acoustical resistance as possible without too much deterioration of the response.

REFERENCES

Bauch, F. W. O. "New High Grade Condenser Microphones," *Journal of the Audio Engineering Society*, Vol. 1, No. 3, p. 232, 1953.

Bauer, B. B. "A Miniature Microphone for Transistorized Amplifiers," *Journal of the Acoustical Society of America*, Vol. 25, No. 5, p. 867, 1953.

Bauer, B. B. "Uniphase Unidirectional Microphones," *Journal of the Acoustical Society of America*, Vol. 13, No. 1, p. 41, 1941.

Bleazey, John C. "Experimental Determination of the Effectiveness of Wind Screens," *Journal of the Audio Engineering Society*, Vol. 9, No. 1, p. 48, 1961.

Carlisle, Richard W. "History and Current Status of Miniature Variable Reluctance and Balanced Armature Transducers," *Journal of the Audio Engineering Society*, Vol. 13, No. 1, p. 45, 1965.

Hanson, O. B. "Microphone Techniques in Broadcasting: Parabolic Reflectors," *Journal of the Acoustical Society of America*, Vol. 3, No. 1, p. 81, 1931.

Inglis, A. H. and Tuffnell, W. L. "An Improved Telephone Set," *Bell System Technical Journal*, Vol. 30, No. 2, p. 239, 1951.

Kaufman, Barry M. "Closing the Wireless Versus Wired Microphone Dependability Gap," *Journal of the Audio Engineering Society*, Vol. 19, No. 1, p. 46, 1971.

Killion, Mead C. and Carlson, Elmer V. "A Wideband Miniature Microphone," *Journal of the Audio Engineering Society*, Vol. 18, No. 6, p. 631, 1970.

Mason, W. P. and Marshal, R. N. "A Tubular Directional Microphone," *Journal of the Acoustical Society of America*, Vol. 10, No. 3, p. 206, 1939.

Olson, Harry F. *Acoustical Engineering*, Van Nostrand Reinhold, New York, 1957.

Olson, Harry F. "A Unidirectional Ribbon Microphone," *Journal of the Acoustical Society of America*, Vol. 3, No. 3, p. 315, 1932.

Olson, Harry F. "Directional Microphones," *Journal of the Audio Engineering Society*, Vol. 14, No. 4, p. 420, 1967.

Olson, Harry F. "Gradient Microphones," *Journal of the Acoustical Society of America*, Vol. 17, No. 3, p. 192, 1946.

Olson, Harry F. "Line Microphones," *Proceedings of the Institute of Radio Engineers,* Vol. 27, No. 7, p. 438, 1939.

Olson, Harry F. "Mass Controlled Electrodynamic Microphones; The Ribbon Microphone," *Journal of the Acoustical Society of America*, Vol. 3, No. 1, p. 695, 1931.

Olson, Harry F. "Polydirectional Microphones," *Proceedings of the Institute of Radio Engineers*, Vol. 32, No. 2, p. 77, 1944.

Olson, Harry F. "Ribbon Velocity Microphones," *Journal of the Audio Engineering Society*, Vol. 18, No. 3, p. 263, 1970.

Olson, Harry F. *Solutions of Engineering Problems by Dynamical Analogies*, Van Nostrand Reinhold, New York, 1966. Provides the means for deriving and solving the dynamical analogies considered in this chapter.

Olson, Harry F. and Preston, John. "Electrostatic Uniangular Microphone," *Journal of the Society of Motion Picture and Television Engineers*, Vol. 67, No. 11, p. 750, 1958.

Olson, Harry F., Preston, John, and Bleazey, John C. "Bigradient Uniaxial Microphone," *RCA Review*, Vol. 17, No. 4, p. 522, 1956.

Olson, Harry F., Preston, John, and Bleazey, John C. "The Uniaxial Microphone," *RCA Review*, Vol. 14, No. 1, p. 47, 1953.

Olson, Harry F. "Microphone Thermal Agitation Noise," *Journal of the Acoustical Society of America*, Vol. 51, No. 2, p. 425, 1972.

Sessler, G. M. and West, J. E. "Condenser Microphones with Electret Foil," *Journal of the Audio Engineering Society*, Vol. 12, No. 2, p. 129, 1964.

Tremaine, Howard M. *Audio Cyclopedia*, Howard W. Sams, Indianapolis, 1969

Wente, E. C. "A Condenser Microphone As a Uniformly Sensitive Instrument For the Absolute Measurement of Sound Intensity," *Physical Review*, Vol. 10, No. 1, p. 39, 1917.

Wente, E. C. and Thuras. "Moving Coil Telephone Receivers and Microphones," *Journal of the Acoustical Society of America*, Vol. 3, No. 1, p. 44, 1931.

(See also numerous articles on microphones in the *Journal of the Audio Engineering Society*.)

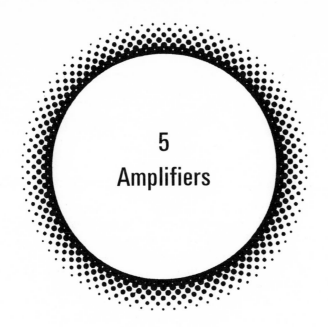

5
Amplifiers

5.1 INTRODUCTION

An audio amplifier is an active electronic transducer in which the output voltage, current, or power corresponds to and usually exceeds the input voltage, current, or power. External power is supplied for the transduction process. The most common audio amplifiers employ vacuum tubes, transistors, and integrated circuits. Transistors and integrated circuits have completely displaced vacuum tubes in modern audio amplifiers. Therefore, the considerations of amplifiers in this chapter will be confined to transistor and integrated-circuit audio amplifiers.

5.2 TRANSISTORS

A. Common Transistors

The transistor is a three-element solid state active electronic transducer used in modern audio amplifiers. There are three elements in the common transistor as depicted in Fig. 5.1. The first two letters in the *n-p-n* and *p-n-p* designations indicate the respective polarities of the voltages applied to the emitter and the collector in normal operation.

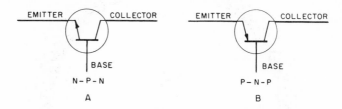

Fig. 5.1 The elements of a transistor. A. *n-p-n* type. B. *p-n-p* type.

In the *n-p-n* transistor shown in Fig. 5.1A, electrons flow from the emitter to the collector. In the *p-n-p* transistor shown in Fig. 5.1B, electrons flow from the collector to the emitter. In other words, the direction of dc electron current is always opposite to that of the arrow on the emitter head. The "conventional current flow" is in the direction of the arrow.

In an *n-p-n* transistor, the emitter is made negative with respect to both the collector and the base, and the collector is made positive with respect to both the emitter and the base. In a *p-n-p* transistor, the emitter is made positive with respect to both the collector and the base, and the collector is made negative with respect to both the emitter and the base.

There are three basic ways of connecting transistors in a circuit: common-base, common-emitter, and common-collector. In the common-base (or grounded-base) connection shown in Fig. 5.2A, the signal is introduced into the emitter-base circuit and extracted from the collector-base circuit. (Thus the base element of the transistor is common to both the input and the output circuits.) Because the input or emitter-base circuit has a low impedance, (resistance plus reactance) in the order of 0.5 to 50 ohms, and the output or collector-base circuit has a high impedance in the order of 1,000 ohms to 1 megohm, the voltage gain is large. The power gain in this type of configuration may be in the order of 1,500.

The direction of the arrows in Fig. 5.2A indicates conventional current flow. As stated previously, most of the current from the emitter flows to the collector; the remainder flows through the base. In practical transistors, from 95 to 99.5 percent of the emitter current reaches the collector. The current gain of this configuration, therefore, is always less than unity, usually in the order of 0.95 to 0.995.

The waveforms in Fig. 5.2A represent the input voltage produced by the signal generator e_s and the output voltage developed across the load resistor r_{EL}. When the input voltage is positive, as shown at AB, it opposes the forward bias produced by the base-emitter battery, and thus reduces current flow through the *n-p-n* transistor. The reduced electron-current flow through r_{EL} then causes the top point of the resistor to become less negative (or more positive) with respect to the lower point, as shown at $A'B'$ on the output

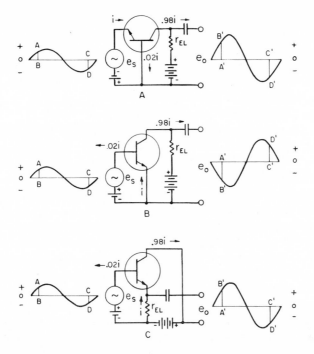

Fig. 5.2   Transistor circuit configurations with input and output waveforms. A. Common base. B. Common emitter. C. Common collector. e_s = signal input voltage. e_o = output voltage. i = current.

waveform. Conversely, when the input signal is negative, as at CD, the output signal is also negative, as at $C'D'$. Thus, the phase of the signal remains unchanged in this circuit; that is, there is no voltage phase reversal between the input and the output of a common-base transistor amplifier.

In the common-emitter (or grounded-emitter) connection shown in Fig. 5.2B, the signal is introduced into the base-emitter circuit and extracted from the collector-emitter circuit. This configuration has more moderate input and output impedances than the common-base circuit. The input (base-emitter) impedance is in the range of 20 to 5000 ohms, and the output (collector-emitter) impedance is about 50 to 50,000 ohms. Power gains in the order of 10,000 (or approximately 40 decibels) can be realized with this circuit because it provides both current gain and voltage gain.

Current gain in the common-emitter configuration is measured between the base and the collector, rather than between the emitter and the collector as in the common-base circuit. Because a very small change in base current produces a relatively large change in collector current, the current gain is always greater than unity in a common-emitter circuit; a typical value is about 50.

The input-signal voltage undergoes a phase reversal of 180 degrees in a common-emitter amplifier, as shown by the waveforms in Fig. 5.2B. When the input voltage is positive, as shown at AB, it increases the forward bias across the base-emitter junction, and thus increases the total current flow through the transistor. The increased electron flow through r_{EL} then causes the output voltage to become negative, as shown at $A'B'$. During the second half-cycle of the waveform, the process is reversed; i.e., when the input signal is negative, the output signal is positive (as shown at CD and $C'D'$).

The third type of connection, shown in Fig. 5.2C, is the common-collector (or grounded-collector) circuit. In this configuration, the signal is introduced into the base-collector circuit and extracted from the emitter-collector circuit. Because the input impedance of the transistor is high and the output impedance low in this connection, the voltage gain is less than unity and the power gain is usually lower than that obtained in either a common-base or a common-emitter circuit. The common-collector circuit is used primarily as an impedance-matching device. As in the case of the common-base circuit, there is no phase reversal of the signal between the input and the output.

The circuits shown in Figs. 5.2A, B, and C are biased for *n-p-n* transistors. When *p-n-p* transistors are used, the polarities of the batteries must be reversed. The voltage phase relationships, however, remain the same.

B. Field Effect Transistors

From the standpoint of audio amplifiers, the field effect transistor also termed MOS (metal-oxide-semiconductor) field effect transistor provides a high input impedance—as a matter of fact, an even higher input impedance than is possible in vacuum tubes.

The field effect transistor uses a metal gate electrode separated from the semiconductor by an insulator as depicted schematically in Fig. 5.3A. The gate, which presents a high input impedance, controls the action of the field effect transistor.

The common-source circuit arrangement shown in Fig. 5.3B is most frequently used in field effect transistor applications. The configuration provides a high input impedance, medium-to-high output impedance, and a voltage gain greater than unity. The voltage gain is of the order of 4 to 200. The input capacitance is from 1 to 5 picofarads. The input resistance is of the order of 10^{14} ohms.

The addition of an unbypassed source resistor to the circuit of 5.3B as shown in Fig. 5.3C produces negative voltage feedback proportional to the

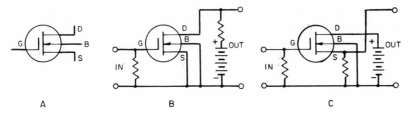

Fig. 5.3   A. Elements of a MOS field effect transistor. G = gate. D = drain. S = source and B = active bulk. B. Basic common-source circuit for a field effect transistor. C. Basic common-drain or source follower for a field effect transistor.

output current. The configuration depicted in Fig. 5.3C is termed the common-drain arrangement. Sometimes the configuration of Fig. 5.3C is called the source follower. In this configuration, the input impedance is higher than in the common-source configuration, the output impedance is low, there is no polarity reversal between input and output, the voltage gain is always less than unity, and distortion is low. The source-follower is used in applications which require reduced input-circuit capacitance, downward impedance transformation, or increased input-signal handling capability. The input signal is effectively injected between gate and drain, and the output is taken between source and drain. The circuit inherently has 100 percent negative voltage feedback. The voltage gain is of the order of 0.6 to 0.8.

5.3 TRANSISTOR AUDIO AMPLIFIERS

A. Audio Amplifier Considerations

The amplifying action of a transistor can be used in various ways in electronic circuits, depending on the results desired. The three recognized classes of audio amplifier service can be defined for transistor circuits as follows: A *class A amplifier* is an amplifier in which the base bias and alternating signal are such that collector current in a transistor flows continuously during the complete electrical cycle of the signal, and even when no signal is present. A *class AB amplifier* is an amplifier in which the base bias and alternating signal are such that collector current in a transistor flows for appreciably more than half but less than the entire electrical cycle. A *class B amplifier* is an amplifier in which the base is biased to approximately collector-current cutoff, so that collector current is approximately zero when no signal is applied, and so that collector current in a transistor flows for approximately one-half of each cycle when an alternating signal is applied.

Audio amplifier circuits are used in radio and television receivers, public address systems, sound recorders and reproducers, and similar applications to amplify signals in the frequency range from 20 to 20,000 hertz. Each transistor in an audio amplifier can be considered as either a current amplifier or a power amplifier. The type of circuit configuration selected is dictated by the requirements of the given application. The output power to be supplied, the required sensitivity and frequency response, and the maximum distortion limits, together with the capabilities and limitations of available devices, are the main criteria used to determine the circuit that will provide the desired performance most efficiently and economically.

In addition to the consideration that must be given to the achievement of performance objectives and the selection of the optimum circuit configuration, the circuit designer must also take steps to insure reliable operation of the audio amplifier under varying conditions of signal level, frequency, ambient temperature, load impedance, line voltage, and other factors which may subject the transistors to either transient or steady state high stress levels.

B. Simple Low-Level Transistor Amplifier

The circuit of a simple low-level transistor amplifier stage is shown in Fig. 5.4. In the circuit configuration shown in Fig. 5.4 the resistor r_{E1} determines the base bias for the transistor. The output signal is developed across the load resistor r_{E2}. The collector voltage and the emitter current are kept relatively low to reduce the noise figure. If the load impedance is low compared to r_{E2}, very little voltage swing results on the collector. Therefore, ac feedback through r_{E1} does not cause much reduction in gain. In the configuration of Fig. 5.4, power gain of up to 10,000 can be obtained. The input impedance is from 20 to 5000 ohms. The output impedance is about 50 to 10,000 ohms. Stages of the type shown in Fig. 5.4 can be used in cascade to obtain the desired overall gain.

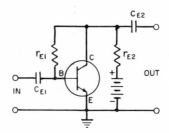

Fig. 5.4   Simple low-level class A transistor amplifier circuit.

C. Bass and Treble Compensation

In sound-reproducing amplifiers means are provided for attenuating or accentuating the low frequency ranges and the high frequency ranges. These means are termed bass and treble compensation. A transistor amplifier with bass and treble controls is shown in Fig. 5.5. The values of the coupling capacitors C_{E0} are so large that the effect on the frequency response is negligible. In the bass boost position, r_{E3} is in parallel with C_{E2} and in series with r_{E1} and r_{E2}. The result is accentuation of the low frequency response. In the bass cut position, r_{E3} is in parallel with C_{E1} and in series with r_{E1} and r_{E2}. The result is attenuation of the low frequency response. In the mid-position the response is uniform with no low frequency accentuation or attenuation. At very high frequencies, C_{E1} and C_{E2} are effectively short circuits and the network is a simple resistance divider r_{E1} and r_{E2}. In the treble boost position, r_{E4} is bypassed by the capacitor C_{E4}, and the result is accentuation of the high frequency response. In the treble cut position, r_{E5} is bypassed by the capacitor C_{E5}, and the result is attentuation of the high frequency response. At very low frequencies the capacitors are of no consequence, and the resistors r_{E4} and r_{E5} act as voltage dividers. In the mid-position the high frequency response is uniform with no high frequency attentuation or accentuation. The overall volume control is provided by the potentiometer r_{E6}.

Fixed frequency response compensation can be carried out by the proper selection of resistance and capacitor networks employing the principles outlined above.

D. Transistor Driver Stage

Driver stages in transistor amplifiers are located immediately before the power-output stage. When a single-ended class A output stage is used, the

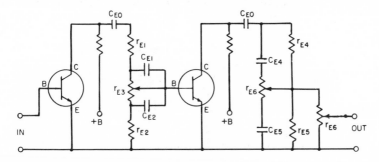

Fig. 5.5 The circuit of a two stage transistor amplifier with separate bass and treble controls and a volume control.

driver stage is similar to a preamplifier stage shown in Fig. 5.4. When a push-pull output stage in which both transistors are the same type (*n-p-n* or *p-n-p*) is used, however, the audio driver must provide two output signals, each 180 degrees out of phase with the other. This phase requirement can be met by use of a tapped-secondary transformer between a single-ended driver stage and the output stage, as shown in Fig. 5.6. The transformer T_1 provides the required out-of-phase input signals for the two transistors Q_1 and Q_2 in the push-pull output stage.

E. Transistor Audio Power Amplifiers

Transistor audio power amplifiers are used to drive loudspeakers, disk recording cutters, optical recording galvanometers, magnetic recording heads, and earphones. The power range is from a few milliwatts for a hearing aid to a kilowatt for large-scale sound reproduction.

Transistor audio power amplifiers may be class A single-ended stage or class A, AB, or B push-pull stage.

A simple class A single-ended power amplifier is shown in Fig. 5.7. Component values which will provide the desired output can be calculated from the transistor characteristics and the supply voltage according to the procedures outlined in manufacturers' manuals. For example, 4 watts can be obtained from a voltage supply of 14.5 volts.

A typical class B push-pull audio amplifier is shown in Fig. 5.8. Resistors r_{E1} and r_{E2} are the emitter stabilizing resistors. Resistors r_{E3} and r_{E4} form a voltage-divider network which provides the bias for the transistors. The base-emitter circuit is biased near collector cutoff so that very little collector power is dissipated under no-signal conditions. The characteristics of the bias network must be very carefully chosen so that the bias voltage will be just sufficient to minimize cross-over distortion at low signal levels. Because the collector current, collector dissipation, and dc operating point of a

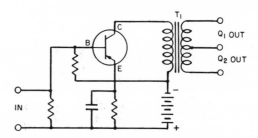

Fig. 5.6 Transistor driver stage for push-pull output circuit.

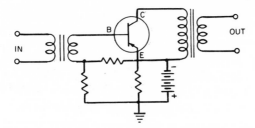

Fig. 5.7   Transistor class A power amplifier circuit.

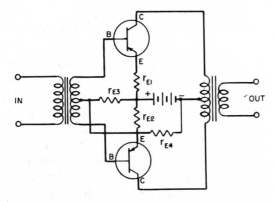

Fig. 5.8   Transistor class B push-pull power amplifier circuit.

transistor vary with ambient temperature, a temperature-sensitive resistor (such as a thermistor) or a bias-compensating diode may be used in the biasing network to minimize the effect of temperature variations.

5.4 DIODES

A. Common Diodes

The simplest type of semiconductor device is the diode, which is represented by the symbol shown in Fig. 5.9A. Structurally, the diode is basically a p–n junction. The n-type material which serves as the negative electrode is referred to as the cathode, and the p-type material which serves as the positive electrode is referred to as the anode. The arrow symbol used for the anode represents the direction of "conventional current flow"; electron current flows in a direction opposite to the arrow.

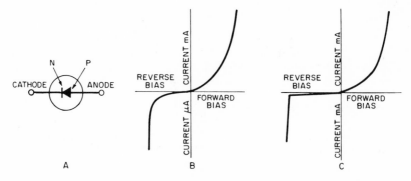

Fig. 5.9 A. Elements of an *n-p* diode. B. Current characteristics of a diode for forward and reverse bias. C. Current characteristics of a Zener diode for forward and reverse bias.

The generalized voltage-current characteristic for a *p–n* junction diode is shown in both the forward and reverse bias directions in Fig. 5.9B. In the forward bias direction the current rises quite rapidly as the voltage is increased and becomes quite large. Current in the reverse direction is usually very low, as depicted in Fig. 5.9B.

One of the most widely used types of semiconductor diode is the silicon rectifier. Silicon rectifier devices are available in a wide range of current capabilities, ranging from tenths of an ampere to several hundred amperes or more, and are capable of operation at voltages as high as 1000 volts or more. Parallel and series arrangements of silicon rectifiers permit even further extension of current and voltage limits.

B. Zener Diodes

Referring to the characteristic of a diode in Fig. 5.9B: when a reverse bias is applied to a silicon rectifier, a limited amount of reverse current, usually measured in microamperes, begins to flow. At a specific reverse voltage which varies for different diodes, a very sharp increase in reverse current occurs. This voltage is called the avalanche or Zener voltage. In many applications, rectifiers can operate safely at the Zener or avalanche point. This voltage is a very definite value and the current increases rapidly with only a small increase in voltage beyond this point, as is shown in Fig. 5.9C. The Zener diode gives a means of providing a fixed voltage when the supply voltage varies over a wide range. For example, if a Zener diode is connected in series with a resistor and a voltage supply, the voltage across the Zener diode will remain constant for large changes in voltage beyond the Zener voltage. Zener diodes operating over a range of 4 to 12 volts are available. Higher voltages may be obtained by cascading two more Zener diodes.

5.5 FEEDBACK AMPLIFIERS

The feedback amplifier consists of an amplifier with means in which a voltage derived from the amplifier output is added to the amplifier input in such a manner as to oppose the applied input signal in the operating range of the amplifier, as is shown in Fig. 5.10. In general, negative feedback applied to an amplifier improves the frequency response and reduces non-linear distortion and reduces some types of noise.

In the system of Fig. 5.10 the ratio of the output and input voltages in the presence of feedback is given by:

$$\frac{e_0}{e_i} = \frac{A}{1 - AB} \tag{5.1}$$

where e_0 = output voltage,

e_i = input voltage,

A = amplification in the absence of feedback, and

B = fraction of the output voltage that is added to input voltage.

In Equation (5.1) the signs are selected so that when the signal voltage is opposed by the feedback voltage, B is negative.

The quantity AB is termed the feedback factor. Equation (5.1) shows that if the feedback factor is large, the amplification is independent of the characteristics of the amplifier. Therefore, under proper conditions of zero phase shift in the feedback circuit, the net effect of feedback is to make the response of the amplifier independent of the frequency.

Feedback reduces the nonlinear distortion produced in the amplifier. The reduction in nonlinear distortion by the application of feedback is given by:

$$\frac{D_F}{D_W} = \frac{1}{1 - AB} \tag{5.2}$$

where D_F = nonlinear distortion with feedback, and

D_W = nonlinear distortion without feedback.

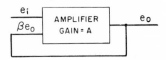

Fig. 5.10   An amplifier with feedback. The amplifier gain without feedback is A. The input voltage to the amplifier = e_i. The output voltage of the amplifier = e_o. β = fraction of the output voltage that is added to the input voltage.

Equation (5.2) shows that a large reduction in nonlinear distortion is obtained by the use of feedback.

Feedback reduces the signal-to-noise ratio in an amplifier. The reduction in signal-to-noise ratio by the application of feedback is given by:

$$\frac{R_F}{R_W} = \frac{A_F}{A_o(1 - AB)} \tag{5.3}$$

where R_F = signal-to-noise ratio with feedback,

R_W = signal-to-noise ratio without feedback,

A_F = amplification between the point where the noise is introduced and the output and with feedback, and

A_o = amplification between the point where the noise is introduced and the output without feedback.

To obtain the advantages of feedback as outlined in the preceding paragraphs, the phase relations must be maintained so that the feedback voltage is 180° out of phase with the input voltage.

5.6 OPERATIONAL AMPLIFIERS

The operational amplifier is a high-gain direct-coupled active network where the gain, frequency response, and impedance are controlled by external networks.

The operational amplifier and the nomenclature associated with it are shown in Fig. 5.11A. A bipolar supply voltage is usually employed. However, a single unipolar supply voltage either positive or negative may be used.

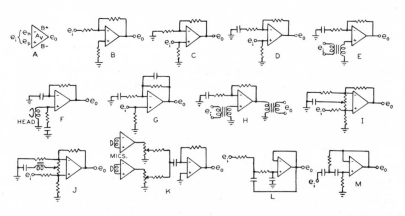

Fig. 5.11   Operational amplifier systems. (After Losmandy.)

The output voltage of the amplifier depicted in Fig. 5.11A is given by:

$$e_o = A_v e_i = (e_n - e_p)A_v \tag{5.4}$$

where
e_o = output voltage,
e_i = input voltage,
A_v = open loop gain of the amplifier,
e_n = voltage input to the inverting section, and
e_p = voltage input to the noninverting section.

The inverting gain operational amplifier network with external feedback is shown in Fig. 5.11B. There is a phase reversal of 180° in this network.

The noninverting gain operational amplifier network with external feedback is shown in Fig. 5.11C. There is no phase reversal in this network.

An audio amplifier operational amplifier network is shown in Fig. 5.11D. In audio amplifiers the response is usually attenuated below the audio range, which is accomplished in this amplifier by the resistor-capacitor combination. There is no phase reversal in this amplifier.

A microphone operational preamplifier network is shown in Fig. 5.11E. The high input impedance makes it possible to use a step up transformer with an output impedance of 50,000 to 100,000 ohms.

An operational amplifier network for a magnetic tape head is shown in Fig. 5.11F.

An operational amplifier network for disk or magnetic tape equalization is shown in Fig. 5.11G. Some modifications may be made in the external network to obtain the desired drop in response with frequency.

A line-to-line operational amplifier network is shown in Fig. 5.11H.

A treble or high frequency control operational amplifier network is shown in Fig. 5.11I. A boost of 20 dB and a cut of 20 dB may be obtained.

A bass or low frequency control operational amplifier is shown in Fig. 5.11J. A boost of 20 dB and a cut of 20 dB may be obtained.

A microphone mixer operational amplifier network is shown in Fig. 5.11K.

A low-pass active filter employing an operational amplifier network is shown in Fig. 5.11L.

A high-pass active filter employing an operational amplifier network is shown in Fig. 5.11M.

The low- and high-pass filters may be used in tandem to obtain a band-pass filter. An external network consisting of an inductance and capacitance in parallel may be used to obtain a narrow band-pass filter.

The preceding brief exposition shows that the operational amplifier is a basic building block from which a large variety of circuit and network functions may be obtained by the addition of appropriate external networks.

5.7 INTEGRATED CIRCUITS

The preceding considerations of audio amplifiers have involved single and separate active elements (transistors) and passive elements (resistors and capacitors) to make up the entire circuit configuration of audio amplifiers. In the integrated circuit the active and passive elements are processed simultaneously from common materials. A typical integrated-circuit audio amplifier will be described in this section.

The elements of an integrated-circuit audio driver amplifier CA3007 are depicted in Fig. 5.12. All the elements shown in Fig. 5.12 are produced on a single flat insulator termed a chip. The input stage consists of a differential pair of transistors Q_1 and Q_2 operating as a phase splitter with gain. The two output signals from the phase splitter, which are 180° out of phase, are direct-coupled through two emitter transistor followers, Q_4 and Q_5, while Q_6 and the associated resistors provide the feedback. The diodes D_1 and D_2 and the transistor Q_3 provide a temperature-stabilizing means, and Q_3 is also a part of the squelch system. When more than 5 volts is applied to terminal 2 the amplifier is immobilized. Transistors Q_4 and Q_5 are intended to drive the pair of power amplifier transistors. The power gain of the integrated circuit CA3007 is 22 dB.

The circuit diagram of the integrated-circuit audio driver amplifier CA3007

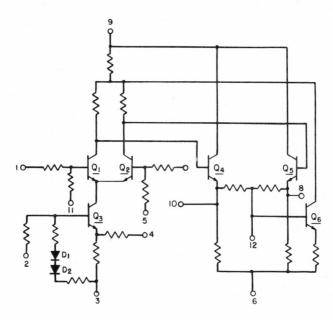

Fig. 5.12 The elements of an integrated circuit audio driver amplifier. RCA No. CA3007.

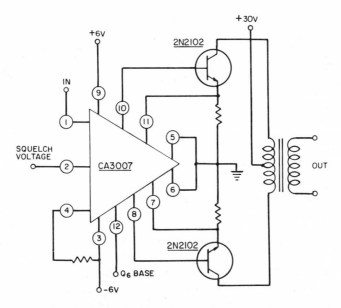

Fig. 5.13 The circuit diagram of the integrated circuit driver amplifier CA3007 connected to two power transistors in a push-push arrangement.

connected to two power transistors is shown in Fig. 5.13. The power output for this configuration is 300 milliwatts.

The integrated-circuit audio driver amplifier CA3007 of Fig. 5.13 is a rather complicated combination of elements and indicates the advantages of the integrated-circuit concept. Audio operational amplifiers and power amplifiers are also available in integrated-circuit form.

5.8 DC POWER SUPPLIES

Batteries and power supplies are used to provide the dc voltages and power requirements in transistor audio amplifiers. A power supply converts an ac power line voltage to a dc voltage suitable for the power and voltage requirements of the transistor audio amplifiers. Power supplies are usually employed when a battery is not adequate or suitable. A power supply consists of a rectifying and ripple smoothing means.

A half-wave rectifier circuit and the voltage output are shown in Fig. 5.14A. The output contains one voltage pulse per cycle. As a consequence, this circuit contains a very high percentage of output ripple.

A full-wave rectifier circuit with a center tapped secondary on the transformer and the voltage output are shown in Fig. 5.14B. The output contains

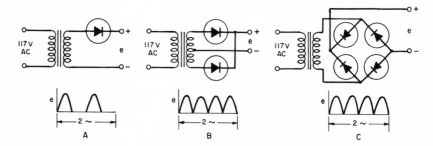

Fig. 5.14 A. Circuit and waveform of a half-wave rectifier. B. Circuit and waveform of a full rectifier with a center tapped transformer. C. Circuit and waveform of a full wave bridge arrangement employing four diodes.

two voltage pulses per cycle. This type of circuit is widely used in powering large audio amplifiers.

A full-wave bridge circuit employing four diodes and the voltage output are shown in Fig. 5.14C. This circuit does not require a center tap on the transformer. The output contains two voltage pulses per cycle. The circuit of Fig. 5.14C can supply twice as much voltage as the circuit of Fig. 5.14B using the same diodes.

The ripple voltage output of the rectifiers is not suitable for the power supply of audio amplifiers, and means must be provided to smooth out the ripple. Typical smoothing filters are shown in Fig. 5.15. A filter with choke input is shown in Fig. 5.15A. A filter with a capacitance input is shown in Fig. 5.15B. Resistance capacitance filters may also be used as shown in Fig. 5.15C and D. Although the resistance filters are not as efficient as the choke filters, the cost is lower. A ripple filter with a Zener diode output is shown in Fig. 5.15E. The voltage output will remain constant for wide variations in the ac line voltage.

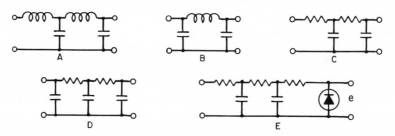

Fig. 5.15 Filter smoothing circuits. A. Choke input. B. Capacitance input. C. Resistance input. D. Capacitance input. E. Zener diode at the output to provide a constant voltage e.

5.9 NOISE

Noise is any undesired audio signal in the sound-reproducing system. In general, noise is an erratic, intermittent, or statistically random oscillation.

One of the important characteristics of a low-level amplifier circuit is the signal-to-noise ratio, or noise figure. The input circuit of an amplifier inherently contains some thermal noise contributed by the resistive elements in the input device. All resistors generate a predictable quantity of noise power as a result of thermal activity.

The voltage generated in a resistor due to the thermal agitation of the electrons is given by:

$$e^2 = 4KTr_E(f_2 - f_1) \tag{5.5}$$

where e = developed voltage, in volts,
K = Boltzmann's constant = 1.374×10^{-23} joule per degree Kelvin,
T = absolute temperature, in degrees Kelvin,
r_E = resistance of the resistor in which the voltage is developed, in ohms, and
$f_2 - f_1$ = frequency bandwidth, in hertz.

The noise power is given by:

$$N = \frac{e^2}{r_E} = 4KT(f_2 - f_1) \tag{5.6}$$

where N = noise power, in watts.

The noise power is -156 dB referenced to one watt for a bandwidth of 15 kilohertz and 20°C. The available power into a matched noiseless load is $e^2/4_E$ or -162 dB. If the reference is 1 milliwatt (which is 0 dBM), the respective levels are -126 dBM and -132 dBM.

When an input signal is amplified, therefore, the thermal noise generated in the input circuit is also amplified. If the ratio of signal power to noise power (S/N) is the same in the output circuit as in the input circuit, the amplifier is considered to be "noiseless" and is said to have a noise figure of unity, or zero dB.

In practical circuits, however, the ratio of signal power to noise power is inevitably impaired during amplification as a result of the generation of additional noise in the circuit elements. A measure of the degree of impairment is called the noise figure of the amplifier, and is expressed as follows:

$$NF = \frac{S_i/N_i}{S_o/N_o} \tag{5.7}$$

where $\quad NF$ = noise figure of the amplifier,

$\quad S_i$ = signal power at the input,

$\quad N_i$ = noise power at the input,

$\quad S_o$ = signal power at the output, and

$\quad N_o$ = noise power at the output.

The noise figure in dB is equal to ten times the logarithm of this power ratio. For example, an amplifier with a 1-dB noise figure decreases the signal-to-noise ratio by a factor of 1.26, a 3-dB noise figure by a factor of 2, a 10-dB noise figure by a factor of 10, and a 20-dB noise figure by a factor of 100.

In audio amplifiers, it is desirable that the noise figure be kept low. In general, the lowest value of NF is obtained by use of an emitter current of less than 1 milliampere and a collector voltage of less than 2 volts for a signal-source resistance between 300 and 3000 ohms. If the input impedance of the transistor is matched to the impedance of the signal source, the lowest value of NF that can be attained is 3 dB. Generally, the best noise figure is obtained by use of a transistor input impedance approximately 1.5 times the source impedance. However, this condition is often not realizable in practice because many transducers are reactive rather than resistive. In addition, other requirements such as circuit gain, signal-handling capability, and reliability may not permit optimization for noise.

The noise level of a high fidelity amplifier determines the range of volume the amplifier is able to reproduce, i.e., the difference (usually expressed in dB) between the loudest and softest sounds in program material. Because the greatest volume range utilized in electrical program material at the present time is about 60 dB, the noise level of a high fidelity amplifier should be at least 60 dB below the signal level at the desired listening level.

5.10 DISTORTION

The main types of distortion in audio amplifiers are frequency discrimination and nonlinear amplitude distortion.

Frequency discrimination is due to a nonuniform frequency response characteristic. There is really no problem in obtaining a uniform response frequency characteristic in an amplifier covering the audio frequency range.

A nonlinear amplitude characteristic leads to the production of spurious harmonics in the reproduced sound. This type of nonlinear distortion is termed harmonic distortion. Harmonic distortion causes a change in the character of an individual tone by the introduction of harmonics not originally present in the program material. For high fidelity reproduction, total harmonic distortion (expressed as a percentage of the output power) should not be greater than about 0.5 percent at the desired listening level.

Intermodulation distortion is a change in the waveform of an individual tone as a result of interaction with another tone present at the same time in the program material. This type of distortion not only alters the character of the modulated tone, but also may result in the generation of spurious signals at frequencies equal to the sum and difference of the interacting frequencies. Intermodulation distortion should be less than 2 percent at the desired listening level. In general, any amplifier which has low intermodulation distortion will have very low harmonic distortion.

The harmonic and intermodulation distortion should be of such a low order of magnitude as to be imperceptible in the reproduced sound. (See Section 17.7.) To achieve this order of performance in an audio amplifier is not a difficult design task. The addition of feedback is a powerful tool in reducing nonlinear distortion and frequency discrimination.

5.11 TRANSIENT RESPONSE

The transient response of an amplifier refers to the faithfulness of response of the amplifier to a sudden change in the electrical input.

The test employed to depict the transient response of an amplifier is the application of a tone burst. The faithfulness of the reproduction of the tone burst is a measure of the transient response of an amplifier. (See Sections 2.5D and 13.10.) In a properly designed amplifier, if the frequency response is smooth, the transient response will be faithful.

REFERENCES

Losmandy, B. J. "Operational Amplifier Applications," *Journal of the Audio Engineering Society*, Vol. 17, No. 1, p. 14, 1969.

RCA. *Linear Integrated Circuits Manual*, IC-42, RCA Commercial Engineering, Harrison, New Jersey, 1970.

RCA. *Transistor Manual*, Technical Series SC-14, RCA Commercial Engineering, Harrison, New Jersey, 1970.

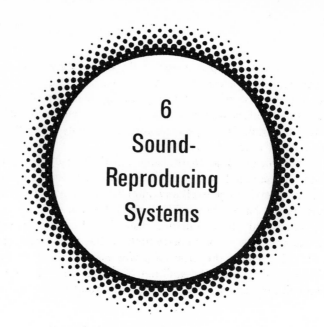

6
Sound-
Reproducing
Systems

6.1 INTRODUCTION

Sound-reproducing systems may be classified as follows: monaural, binaural, monophonic, stereophonic, and quadraphonic. Each of the sound-reproducing systems provides the functions for the particular application. For example, the purpose may be the reproduction of intelligible speech, the reproduction of realism, or the reproduction of emotionalism. However, as the term implies, the main purpose of sound reproduction is to provide realism.

To achieve realism in a sound-reproducing system, four fundamental conditions must be satisfied, as follows:

(1) The frequency range must be such as to include without frequency discrimination all of the audible components of the various sounds to be reproduced.

(2) The volume range must be such as to permit noiseless and distortionless reproduction of the entire range of intensity associated with the sounds.

(3) The reverberation characteristics of the original sound should be approximated in the reproduced sound.

(4) The spatial sound pattern of the original sound should be preserved in the reproduced sound.

The purpose of this chapter is to describe the five types of sound-reproducing systems, the quality of the performance of each system in relation to achieving realism in the reproduction of the sound, and the transmission of information in each system.

6.2 MONAURAL SOUND-REPRODUCING SYSTEM

A schematic diagram depicting the elements of a monaural sound-reproducing system is shown in Fig. 6.1. A monaural sound-reproducing

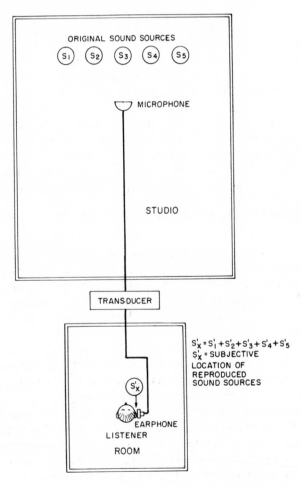

Fig. 6.1   Monaural sound reproducing system.

system is of the closed-circuit type in which one or more microphones, used to pick up the sound, are connected to a single transducing channel which in turn is coupled to one or two earphones worn by the listener.

The most common example of a monaural sound-reproducing system is the telephone, in which there is, in general, a single source of sound, one microphone, usually termed a transmitter, a transducer, and one earphone, usually termed a telephone receiver, coupled to one ear of the listener. In most local applications, the carbon microphone is coupled directly to the earphone. In long distance telephony, vacuum-tube and transistor amplifiers may be used between the microphone and telephone receiver. For other more limited applications, as, for example, monitoring purposes, the transducer may be a radio transmitter and receiver, a television sound transmitter and receiver, a disc phonograph recorder and reproducer, a sound motion picture recorder and a reproducer, and/or a magnetic tape recorder and reproducer. In some applications, there may be more than one sound source. One or more microphones may be used. In some applications two earphones may be used, transmitting the same program to each of the ears of the listener. The acoustics of a single room are involved in the reproduction of the sound, namely, the studio in which the microphone is located. The monaural sound-reproducing system may be constructed so as to satisfy the above conditions (1) and (2) on realism of sound reproduction. Condition (3) on realism is only partially satisfied. The system cannot satisfy condition (4) on realism.

6.3 BINAURAL SOUND-REPRODUCING SYSTEM

A schematic diagram depicting the elements of a binaural sound reproducing system is shown in Fig. 6.2. A binaural sound reproducing system is of the closed-circuit type in which two microphones used to pick up the original sound are each connected to two independent corresponding transducing channels coupled to independent corresponding earphones worn by the listener.

There is no widespread use of the binaural sound-reproducing system. The use is limited to specific applications. The binaural sound-reproducing system consists of two separate channels. Each channel consists of a microphone, transducer, and earphone. The microphones are mounted in a dummy simulating the human head in shape and dimensions and at the locations corresponding to the ears of the human head. The transducer may be an amplifier, a radio transmitter and receiver, a phonograph recorder and reproducer, a motion picture recorder and reproducer, and/or a magnetic tape recorder and reproducer. The listener is transferred to the location

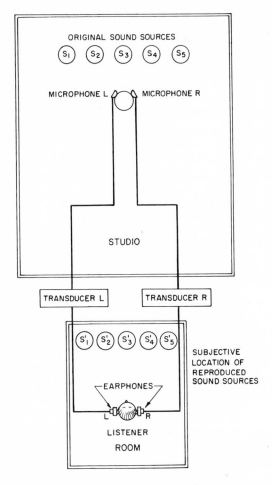

Fig. 6.2 Binaural sound reproducing system.

of the dummy by means of a two-channel sound-reproducing system. The binaural sound-reproducing system may be constructed so as to satisfy all four conditions on realism of sound reproduction.

6.4 MONOPHONIC SOUND-REPRODUCING SYSTEM

A schematic diagram depicting the elements of a monophonic sound-reproducing system is shown in Fig. 6.3. A monophonic sound-reproducing system is of the field type in which one or more microphones, used to pick

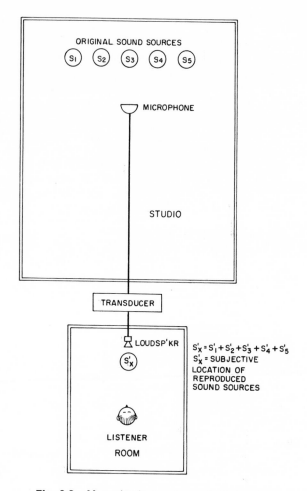

Fig. 6.3 Monophonic sound reproducing system.

up the original sound, are coupled to a single transducing channel which in turn is coupled to one or more loudspeakers in reproduction.

The monophonic sound-reproducing system is more widely employed than any of the other sound-reproducing systems.

Examples of monophonic sound-reproducing systems are the disc phonograph, radio, sound motion picture, television, magnetic tape reproducer, and sound systems. The monophonic sound-reproducing system is of the field type, where sound is picked up by a microphone and reproduced by a loudspeaker into a field. The sound at the microphone is reproduced at the loudspeaker. The transducer may be an amplifier, radio transmitter and receiver, a phonograph recorder and reproducer, a sound motion picture recorder and reproducer, a television transmitter and receiver, and/or a magnetic tape recorder and reproducer. The monophonic sound reproducer may be constructed to satisfy conditions (1) and (2) on realism of sound reproduction. Condition (3) on realism is only partially satisfied. The system does not satisfy condition (4) on realism.

6.5 STEREOPHONIC SOUND-REPRODUCING SYSTEM

A schematic diagram depicting the elements of a stereophonic sound reproducing system is shown in Fig. 6.4. A stereophonic sound-reproducing system

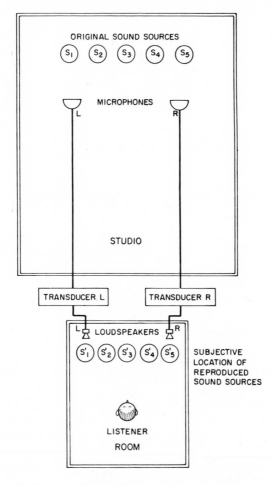

Fig. 6.4 Stereophonic sound reproducing system arranged to provide auditory perspective.

is of the field type in which two or more microphones, used to pick up the original sound, are each coupled to a corresponding number of independent transducing channels which in turn are each coupled to a corresponding number of loudspeakers arranged in substantial geometrical correspondence to the microphones.

The main feature of the stereophonic sound-reproducing system is the reproduction of sound in auditory perspective. The subjective location of the reproduced sound sources corresponds to the location of the original sound sources in direct listening under geometrically corresponding conditions, as depicted in Fig. 6.4.

The transducer may be an amplifier, radio transmitter and receiver, a phonograph recorder and reproducer, a sound motion picture recorder and reproducer, and/or a magnetic tape recorder and reproducer. Two channels are used in the disc phonograph and radio. Two and three channels are used in the magnetic tape reproducer. Two, three, and more channels are used in sound motion picture reproducers. The stereophonic sound reproducer may be constructed to satisfy conditions (1), (2), and (4) on realism of sound reproduction. Condition (3) on realism is only partially satisfied.

6.6 QUADRAPHONIC SOUND-REPRODUCING SYSTEM

A schematic diagram depicting the elements of a quadraphonic sound-reproducing system is shown in Fig. 6.5. A quadraphonic sound-reproducing system is of the field type in which four microphones each coupled to four transducing channels are each coupled to four loudspeakers arranged in substantial geometrical correspondence to the microphones.

At the present time the only commercial quadraphonic transducing system is the four-channel magnetic tape recorder and reproducer.

The quadraphonic sound-reproducing system satisfies conditions (1), (2), (3), and (4) on realism of sound reproduction. Channels left front and right front provide the auditory perspective. Channels left rear and right rear provide the reverberation envelope. Therefore, the quadraphonic system simulates direct listening. For example, in a concert hall or theater, in direct listening there are two major effects, namely, spatial effects and the reverberation envelope. The listener localizes the different sound sources and attributes the proper location to the sound sources. The listener also hears the multireflected sound as a reverberation envelope. Consequently, the quadraphonic sound-reproducing system provides realism in all respects.

The quadraphonic sound-reproducing system may also be employed to produce all manner of acoustic spatial effects by means of the arrangement depicted in Fig. 6.6. The source of sound may be made to move clockwise

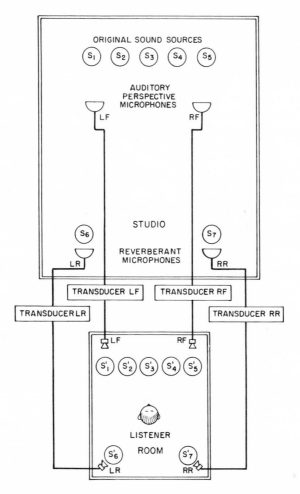

Fig. 6.5 Quadraphonic sound reproducing system arranged to provide auditory perspective and the reverberation envelope.

around the room by suitable switching and fading to produce the reproduced sound sources in the sequence of S'_1 to S'_2 to S'_3, and so on; or, in a counterclockwise direction, by suitable switching and fading to produce the sound sources in sequence of S'_1 to S'_8 to S'_7, and so on. The sound source may be made to bounce back and forth by switching and fading the sound sources in the sequence of S'_3 to S'_7 then S'_5 to S'_1, and so on. In the example of Fig. 6.6, a single original sound source is used. If there are several original sources the results will be very complex. The quadraphonic system provides the means for highly complex and artistic sound effects.

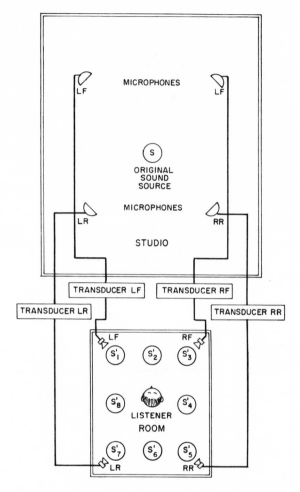

Fig. 6.6 Quadraphonic sound reproducing system arranged to produce sound in motion.

6.7 TRANSMISSION OF INFORMATION

An important consideration in a sound-reproducing system is the capacity of a system for the transmission of information. Employing the best coding, a sound-reproducing system can transmit binary digits (bits) at a rate given by:

$$C = W \log_2 \frac{P + N}{N} \tag{6.1}$$

where C = capacity of the system, in bits per second,
$\quad\quad W$ = frequency bandwidth, in hertz,
$\quad\quad P$ = average power used in the transmission, in ergs per second, and
$\quad\quad N$ = average noise power, in ergs per second.

Table 6.1, employing Equation (6.1), shows the number of bits transmitted per second by one-, two-, and four-channel systems for 60-dB, 50-dB, and 40-dB signal-to-noise ratios and bandwidths of 5000, 10,000, and 15,000 hertz. For the same signal-to-noise ratio, the four-channel system can transmit four times as much information as the one-channel system and two times as much information as the two-channel system. The two-channel system can transmit two times as much information as the one-channel system. Table 6.1 also shows that the amount of information that can be transmitted increases as the signal-to-noise ratio increases. In addition, Table 6.1 shows that the amount of information that can be transmitted increases as the bandwidth increases. The conclusion is that a four-channel system with a 60-dB signal-to-noise ratio and a bandwidth of 15,000 hertz provides the highest capacity for the transmission of information.

TABLE 6.1

Information Transmission Rates of One-, Two-, and Four-Channel Sound Systems for Signal-to-Noise Ratios of 40, 50, and 60 Decibels

Frequency Range in Hz	Signal-to-Noise Ratio in dB	Transmission Rate in bits per second		
		One-Channel	Two-Channel	Four-Channel
5,000	40	66,500	133,000	266,000
5,000	50	83,000	166,000	332,000
5,000	60	99,500	199,000	398,000
10,000	40	133,000	266,000	532,000
10,000	50	166,000	332,000	664,000
10,000	60	199,000	398,000	796,000
15,000	40	199,000	398,000	796,000
15,000	50	250,000	500,000	1,000,000
15,000	60	298,000	597,000	1,194,000

REFERENCE

Olson, Harry F. "Sound Reproducing Systems—Monaural, Binaural, Monophonic and Stereophonic," *Audio*, Vol. 42, No. 9, p. 58, 1958.

7
Telephone, Hearing
Aid, and Sound
Systems

7.1 INTRODUCTION

A simple sound-reproducing system consists of a microphone, an amplifier, and an earphone or a loudspeaker. Examples of simple sound-reproducing systems are the telephone, hearing aids, and sound systems. The telephone is the oldest sound-reproducing system. The telephone instrument combined with complex office and interconnecting means makes it possible for any person to talk to any other person anywhere in the United States in a matter of seconds. Tests made upon a representative cross-section of the people in the United States show a very large percentage to be hard of hearing. Practically all of these people may obtain improved hearing performance from a hearing aid. A hearing aid is a complete sound-reproducing system which increases sound pressure over that normally received by the ear. Sound systems are used for sound-reenforcement, intercommunicating, public address, announce, and paging systems, in theaters, churches, auditoriums, outdoor theaters, mass meetings, athletic events, factories, offices, railroad stations, airports, hotels, hospitals, etc.

7.2 TELEPHONE

The telephone is a sound-reproducing system consisting of a carbon microphone (sometimes termed a transmitter), an earphone (sometimes termed a telephone receiver), and a battery, as shown in Fig. 7.1. The modern carbon microphone used in telephony has been described in Section 4.3. The modern earphone used in telephony has been described in Section 3.2. A very simplified schematic diagram of a telephone system is depicted in Fig. 7.1. The ringing and other elements of the system not connected with sound reproduction have been omitted. A modern subscriber rotary dial telephone set is shown in Fig. 7.2A. A modern subscriber keyboard "touchtone" telephone set is shown in Fig. 7.2B. The battery, ringing generators, switching, and signal equipment are located in the central office. The function of the central office is to connect any subscriber to any other subscriber by manual or automatic means. In large cities, there are many central offices because it is not economical or practical for a central office to serve more than 10,000 subscriber stations. The local offices are interconnected by lines as shown in Fig. 7.3. In local transmission, the electrical output of the carbon microphone fed to the input of the earphone is sufficient for the earphone to generate sound of ample loudness for the intelligent transmission of speech. In long distance

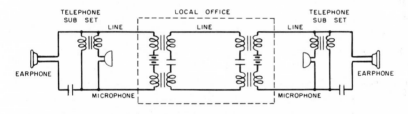

Fig. 7.1   Schematic diagram of the apparatus in a telephone system.

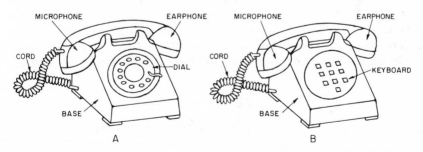

Fig. 7.2   Perspective views of subscriber telephone sets. A. Rotary dial type. B. Keyboard "touchtone" type.

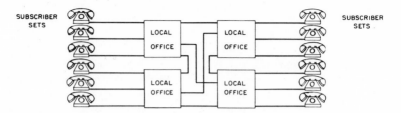

Fig. 7.3 Schematic diagram of a telephone system with four central offices and interconnecting lines.

telephony, vacuum-tube and transistor repeaters are used at regular intervals to restore the level of transmission of the speech signals to a normal value. The systems shown in Figs. 7.1 and 7.3 depict the electroacoustic elements of the telephone system. A consideration of circuits, networks, manual and automatic exchanges, repeaters, etc., is outside of the scope of this book. For information relating to these subjects the reader is referred to books on telephone systems.

7.3 HEARING AID

A hearing aid consists of a microphone, amplifier, volume and tone controls, and an earphone, comprising a sound-reproducing system powered by a battery as shown in Fig. 7.4. The advent of the transistor made it possible to provide a hearing aid of small physical size and low battery power consumption. The microphones used in hearing aids are small versions of the magnetic microphone described in Section 4.3E and the ceramic microphone described in Section 4.3C. The amplifiers used in hearing aids are of the transistor or integrated-circuit types described in Chapter 5. Two types of earphone are used, namely, the insert type described in Section 3.2A and the bone-conduction earphone described in Section 3.6. In some cases a tube is connected between the ear insert and the earphone. In one type of deafness, the middle ear, which consists of bones that conduct sound to the inner ear, is damaged. (See Section 17.2 for a description of the ear.) Under these conditions sound may be transmitted to the inner ear by means of the bone-

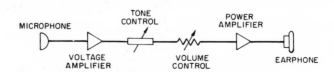

Fig. 7.4 Schematic diagram of a hearing aid.

conduction earphone described in Section 3.6. The face of the bone-conduction earphone is placed against the skin and flesh covering the mastoid bone back of the external ear.

Complete hearing aids are shown in Fig. 7.5. The most common hearing aid consists of a case housing the microphone, amplifier, volume and tone controls and battery. A cord connects the output of the amplifier in the case and the ear insert earphone as shown in Fig. 7.5A. The case is worn at some convenient location on the clothing. Tremendous sound levels can be obtained with the hearing aid of Fig. 7.5A before acoustic feedback oscillation sets in.

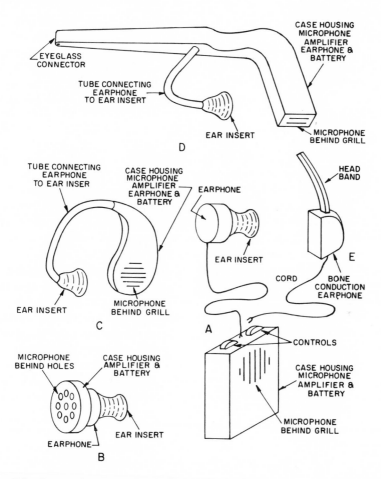

Fig. 7.5   Perspective views of typical hearing aids. A. Body type with insert earphone. B. In the ear type. C. Behind the ear type. D. Eyeglass type. E. Body type with bone conduction earphone.

Therefore, this hearing aid is used for the very hard-of-hearing. A complete hearing aid in a simple compact form is shown in Fig. 7.5B. The entire system, consisting of the microphone, amplifier, volume control, battery, and earphone, is housed in a small case. Because of the close proximity of the microphone and earphone, the sound amplification that can be obtained without acoustic feedback oscillation is somewhat limited. Therefore, the seal between the ear canal and the insert must be very good in order to reduce the effect of feedback between the microphone and the earphone. Another compact form of hearing aid is shown in Fig. 7.5C. The case is worn behind the ear. A tube connects the earphone transducer to the ear insert. The tube is carried over the ear and acts as support for the case. The microphone, amplifier, volume and tone controls, and batteries are housed in the case behind the ear. Since the microphone and earphone are separated at a larger distance than in the hearing aid of Fig. 7.5B, the sound amplification that can be obtained before acoustic feedback oscillation sets in is greater. In another form of hearing aid, the entire system, consisting of the microphone, amplifier, volume control, battery, and earphone, is housed in one of the side pieces of eye spectacles, as shown in Fig. 7.5D. The air-type earphone housed in the side piece is connected by means of a tube to the ear insert. The performance is similar to that of the hearing aid of Fig. 7.5C. The bone-conduction earphone described in Section 3.6 may be incorporated into the earpiece so that the active part of the bone-conduction earphone is in contact with skin and flesh covering the mastoid bone behind the ear. Another form of bone-conduction hearing aid is shown in Fig. 7.5E. The case houses the microphone, amplifier, volume and tone controls, and battery. This system is the same as that used for the ear-insert earphone. A cord connects the output of the amplifier in the case to the bone-conduction earphone. A headband may be used to hold the active part of the bone-conduction receiver in contact with skin and flesh covering the mastoid bone behind the ear as shown in Fig. 7.5E. The bone-conduction earphone may also be stuck to the skin by means of an adhesive, thereby eliminating the headband.

7.4 SOUND SYSTEMS

A sound system, in the form of a sound-reenforcement system in the simplest form, consists of microphone, amplifier, volume and tone controls, and loudspeaker. The sound-reenforcement system is used to augment the output of a speaker, singer, or musical instrument. The sound-reenforcement system will render intelligible speech which before was either too weak to be heard above the ambient noise or too reverberant. In large theaters or halls or outdoors, the volume range of the orchestra at the listener's location is inadequate for full artistic appeal. For that application, the use of a sound-reenforcement system is required for augmenting the intensity of the original sound.

A schematic diagram depicting the elements of a sound-reenforcement system is shown in Fig. 7.6. The microphones used in sound-reenforcement systems are nondirectional, bidirectional, and unidirectional. (See Chapter 4.) The particular applications involving the directional pattern of microphones will be described in Chapter 16.

The amplifiers used in modern sound-reenforcement systems are of the transistor or integrated-circuit types described in Chapter 5. The tone and volume controls used in sound-reenforcement systems were described in Section 5.3C. The loudspeakers used in sound reenforcement systems were described in Chapter 2. The particular applications involving the various types of loudspeakers will be described in Chapter 16.

A completely integrated sound-reenforcement system is the portable electronic megaphone shown in Fig. 7.7. The basic elements of the electronic megaphone are depicted in Fig. 7.6. The battery power supply is also contained in the megaphone. The electronic megaphone produces a sound power level of 100 times that of the human voice before oscillations due to acoustic feedback between the loudspeaker and microphone sets in. This means that the distance over which speech can be understood is 10 times that for a person talking without any aid. The electronic megaphone is used for all manner of sound-reenforcement applications requiring a portable means for augmenting the human voice.

A simple form of sound-reenforcement system is shown in Fig. 7.8. Five different types of microphone mountings are shown in Fig. 7.8, namely, the hand-held, the floor-stand, the overhead-supported, the table-stand, and the lavalier. The hand-held microphone is of the nondirectional or unidirectional type. The table, overhead, and floor-stand microphones are of the nondirectional, bidirectional, and unidirectional types. The lavalier supported by a cord around the neck is usually of the nondirectional type. The amplifier is equipped with five inputs and volume controls on each input thereby selecting and controlling the output from the microphones. There is a master control for setting the overall level. The output of the amplifier is fed to a loudspeaker. The direct radiator loudspeaker housed in a small enclosure is used for small or low-level sound-reenforcement systems. For the hand-held

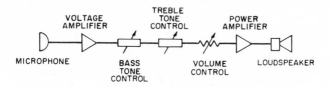

Fig. 7.6 Schematic diagram of a sound reenforcement system.

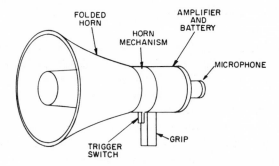

Fig. 7.7 Perspective view of a portable electronic megaphone.

or lavalier microphones the cord or cable presents a cumbersome problem and reduces the mobility of the performer. The radio or "wireless" microphone described in Section 4.8B may also be employed, thereby eliminating the microphone cable or cord.

The simple sound-reenforcement system of Fig. 7.8 depicts a single loudspeaker. Several loudspeakers distributed around the room may be used. Further considerations of sound-reenforcement and announce systems will be carried out in Chapter 16.

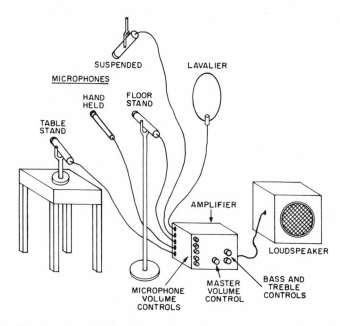

Fig. 7.8 The elements of a small sound reenforcement system with five different types of a microphone mounting.

7.5 INTERCOMMUNICATING SYSTEM

A sound-intercommunicating system is an extension of the simple sound system of Fig. 7.6. A schematic diagram of a sound-intercommunicating system is shown in Fig. 7.9A. A perspective view of the apparatus in a sound-intercommunicating system is shown in Fig. 7.9B. The master unit contains an amplifier, microphone-loudspeaker, station selector, and talk-listen switch. A small loudspeaker with appropriate frequency response compensation also serves as the microphone. The remote unit consists of a microphone-loudspeaker and appropriate switch systems. In the system shown in Fig. 7.9 the remote units can communicate with the master station but not with another remote unit. If all units of the master type and interconnected wiring to all stations are employed, then any station can call and communicate with any other station.

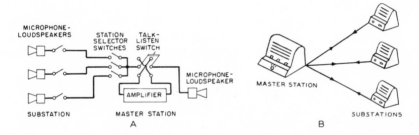

Fig. 7.9 A. Schematic diagrams of the elements in an intercommunicating system. B. Perspective view of the apparatus in an intercommunicating system.

REFERENCES

Burkhard, Mahlon D. "Protection Against Shock and Vibration," *Journal of the Audio Engineering Society*, Vol. 14, No. 1, p. 32, 1966.

Carlisle, Richard W. "History and Current Status of Miniature Variable Reluctance and Balanced Armature Transducers," *Journal of the Audio Engineering Society*, Vol. 13, No. 1, p. 45, 1965.

Inglis, A. H. and Tuffnell, W. L. "An Improved Telephone Set," *Bell System Technical Journal*, Vol. 30, No. 2, p. 239, 1951.

Killion, Mead C. and Carlson, Elmer V. "A Wideband Miniature Microphone," *Journal of the Audio Engineering Society*, Vol. 18, No. 6, p. 631, 1970.

(See also more than fifty articles on sound-reenforcement systems in the *Journal of the Audio Engineering Society* over the past two decades.)

8
Magnetic
Tape Recorders
and Reproducers

8.1 INTRODUCTION

A magnetic recorder consists of an electromagnetic transducer and means for moving a ferromagnetic recording medium relative to the transducer for recording electric signals as magnetic variations in the medium. A magnetic reproducer consists of an electromagnetic transducer and means for moving a ferromagnetic recording medium, with recorded magnetic variations, relative to the transducer to thereby generate voltages in the transducer corresponding to the magnetic variations. The ferroelectric recording medium used in audio recording consists of a plastic base with a coating of magnetic oxide. The purpose of this chapter is to describe the process and the machines employed in magnetic tape sound recording and reproducing of audio signals.

8.2 MAGNETIC TAPE SOUND RECORDING AND REPRODUCING PROCESS

The magnetic tape employed in the recording and reproducing of audio signals consists of a plastic base with a coating of magnetic oxide as depicted in Fig. 8.1A. The base of polyvinylchloride or cellulose acetate used for magnetic tape ranges in thickness from 0.0005 to 0.0015 inch. The magnetic

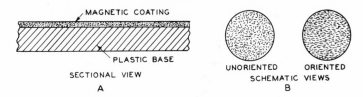

SECTIONAL VIEW
A

UNORIENTED   ORIENTED
SCHEMATIC VIEWS
B

Fig. 8.1 A. Sectional view of magnetic recording tape. B. Schematic views of unoriented and oriented magnetic particles in the tape coating.

coating is applied to the plastic base with the magnetic particles suspended in a liquid air drying and setting plastic formulation. The magnetic coating, when dried and set by the application of heat, ranges in thickness from 0.0002 to 0.0007 inch.

A typical $B - H$ characteristic of the magnetic tape coating is shown in Fig. 8.2; H is the magnetizing force, in oersteds, and B is the magnetic induction, in gausses. The retentivity, B_r, is of the order of 700 to 1600 gausses. The coercivity, H_c, is of the order of 250 to 500 oersteds. The magnetic oxides used for magnetic tape coatings include gamma ferric, cobalt ferric, and chromium dioxide.

In coating the magnetic tape the individual particles are magnetically

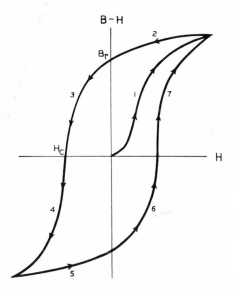

Fig. 8.2 Typical B-H characteristic of the iron oxide coating on magnetic tape. The arrows and numbers depict the tracing of the B-H characteristic. H = magnetizing force. B = magnetic induction. H_c = coercivity. B_r = retentivity.

oriented in the same direction by the application of a magnetic field with the result as shown in Fig. 8.1B. The net effect of orientation is an increase in the retentivity or output of 2 to 3 dB.

The recording and reproducing processes are depicted in Fig. 8.3. The passage of the tape past the recording head leaves a series of magnetized sections which correspond to the signal which was applied to the head when the tape was in contact with the head at each of these sections. In the reproduction process, the tape is moved past the head with the result that a change in magnetic flux is produced in the head as a magnetized section passes the head. This change in flux induces a voltage in the coil which corresponds to the voltage of the original applied signal.

A schematic view and the magnetic network of a magnetic recording head and magnetic tape are shown in Fig. 8.4. The currents i_1 and i_1' produce a flux ϕ in the tape. Because of the retentivity of the coating material of the tape, a magnetized section is produced. The action and performance of the recording process can be deduced from the magnetic network of Fig. 8.4.

A schematic view and magnetic network of a magnetic reproducing head and magnetic tape is shown in Fig. 8.5. When the magnetic tape is moved past the head, the magnetomotive force of the magnetized sections produces the magnetic flux, ϕ_5 and ϕ_5', in the two coils. When this flux changes, as the magnetomotive force changes as different magnetized sections of the tape pass over the head, a voltage, e_1 and e_1', is induced in the coils. These voltages correspond to the currents, i_1 and i_1', applied to the recording head. The magnetic circuit of the magnetic head consists of a stack of laminations of a high-permeability alloy of nickel and iron.

In the recording and reproducing processes, there are losses due to the finite gap length. In the recording process, the gap width is of little importance because the recording process takes place from the trailing edge of the gap. It is this edge rather than the gap that is of importance in the recording head.

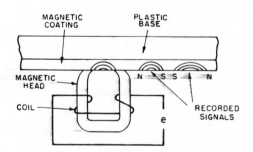

Fig. 8.3 Schematic diagram depicting the magnetic tape recording and reproducing process.

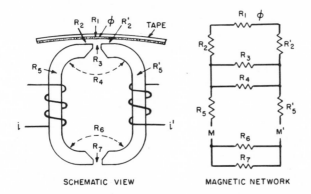

SCHEMATIC VIEW   MAGNETIC NETWORK

Fig. 8.4 Schematic view and magnetic network of the head and tape of a magnetic tape recording system. In the magnetic circuit, M and M' = magnetomotive forces developed by the currents in the two coils. R_5 and R'_5 = reluctances of the magnetic material of the heads. R_3 and R_7 = reluctances of the top and bottom air gaps. R_4 and R_6 = reluctances of the magnetic leakage path. R_2 and R'_2 = reluctances of the air gaps between the head and the tape. R_1 = reluctance of the magnetic tape. ϕ = flux through the magnetic tape. i and i' = currents in the two coils. $M = 4\pi Ni$. N = number of turns on the coil.

In the reproducing process, it is the length of the gap that determines the magnetomotive force.

The output of the reproducing head is given by:

$$e = N\frac{d\phi}{dt} \tag{8.1}$$

where e = voltage output, in abvolts,
                   N = number of turns in the coil,
                   ϕ = flux in the coil, in gausses, and
                   t = time, in seconds.

If the amplitude of ϕ is a constant, the voltage will increase at the rate of 6 dB per octave. However, this characteristic must be multiplied by the gap loss. The open-circuit voltage response of a magnetic reproducing system is given by:

$$R_M = 20\log\left(\sin\frac{\pi d}{\lambda}\right) \tag{8.2}$$

where R_M = open-circuit voltage response, in decibels,
            d = length of the gap, in centimeters, and
            λ = wavelength of the signal along the tape, in centimeters.

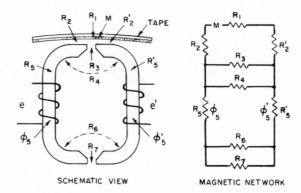

SCHEMATIC VIEW   MAGNETIC NETWORK

Fig. 8.5   Schematic view and magnetic network of the head and tape of a magnetic tape sound reproducing system. In the magnetic circuit, M = magnetomotive force stored in the tape. R_5 and R_5' = reluctances of the magnetic material of the heads. R_3 and R_7 = reluctances of the top and bottom air gaps. R_4 and R_6 = reluctances of the magnetic leakage paths. R_2 and R_2' = reluctances of the air gaps between the head and the tape. R_1 = reluctance of the magnetic tape. ϕ_5 and ϕ_5' = flux in the coils. e and e' = induced voltages in the coils.

The open-circuit voltage response frequency characteristic of a magnetic tape reproducing system is shown in Fig. 8.6. The air gaps in magnetic tape reproducing heads are approximately 0.5, 0.25, 0.12, 0.06, and 0.03 mils for magnetic tape speeds of 30, 15, $7\frac{1}{2}$, $3\frac{3}{4}$, and $1\frac{7}{8}$ inches per second.

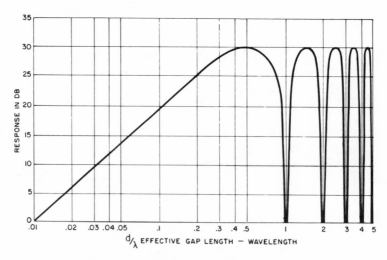

Fig. 8.6   The response of a magnetic tape reproducing system as a function of the ratio of the gap length to the wavelength.

The magnetic material used in the coating of the plastic tape is of necessity nonlinear because it must possess retentivity to retain the magnetic signal applied to the tape in recording. The characteristic which depicts the magnetomotive force or magnetizing force H produced by the recording head in the magnetic tape and the residual induction B_r after the magnetic tape leaves the head is depicted by the characteristic 1, 2, 0, 3, 4 of Fig. 8.7. The nonlinear portion in the vicinity of 0 and in the region of 1 and 4 of Fig. 8.7 will produce nonlinear distortion. Various means have been developed for reducing the effects of this nonlinear characteristic. The system which is universally used in sound reproduction is the high frequency alternating current bias. The high frequency signal of 50 to 150 kilohertz is added to the

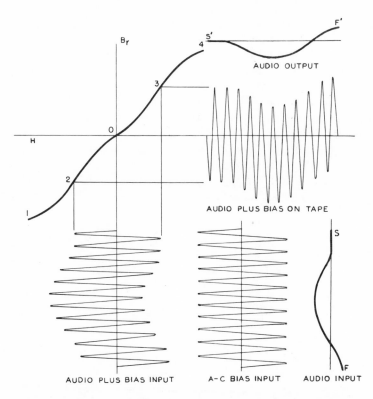

Fig. 8.7 The recording on magnetic tape with a high frequency alternating current bias. The characteristic 1, 2, 0, 3, 4 is the residual induction B_r induced by the magnetizing force H produced by the recording head. The high frequency alternating current bias and the input audio signal S-F are added and fed to the recording head which applies the corresponding magnetomotive force to the tape. The resultant output characteristic in reproduction is the audio signal S'-F'.

audio signal as shown in Fig. 8.7, the high frequency bias is supplied by a transistor oscillator, and the composite signal is applied to the recording head. The action of the high frequency bias in reducing the effect of the nonlinear characteristics of 1, 2, 0, 3, 4 is shown in the output composite signal of Fig. 8.7. Bias can be adjusted to suit the tape in use. In reproducing, the high frequency signal is not reproduced, and the frequency response of the reproducer output is the audio signal shown in Fig. 8.7; it is the same as the input signal.

The system used in recording and reproducing from magnetic tape is shown in Fig. 8.8. The tape transport mechanism consists of the takeup and payoff reels and the capstan drive. The capstan is connected to a large rotational inertia to provide a relatively constant rotational velocity. A rubber pressure roller presses the tape against the capstan to insure good contact of the tape with the capstan. The takeup reel mechanism produces a relatively constant tension in the takeup tape. Some sort of braking on the payoff reel leads to a relatively constant tension in the payoff tape.

In professional and high quality magnetic tape machines separate motors are provided for the capstan, takeup reel, and payoff reel drives as shown in Fig. 8.9. The capstan drive consists of a two-speed synchronous motor with the capstan as an extension of the shaft. In recording or playing the tape the takeup reel motor supplies the appropriate torque to insure relatively constant tension in the takeup tape. The payoff reel motor supplies a braking torque to insure relatively constant tension in the payoff tape. In rewinding, the pressure roller does not engage the tape, and the functions of the takeup and payoff motors are reversed.

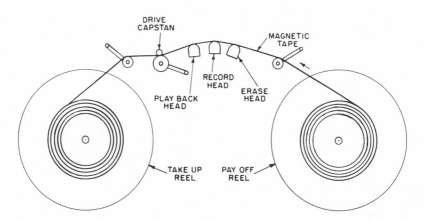

Fig. 8.8   The elements of the transport mechanism of a tape recording and reproducing mechanism.

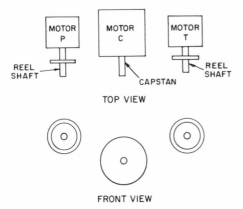

TOP VIEW

FRONT VIEW

Fig. 8.9   The elements of the drive mechanism of a magnetic tape machine in which there is a separate motor for driving the capstan, takeup reel and payoff reel.

In the lower-cost magnetic tape machines a single motor is used to supply the rotational functions of the capstan, takeup, and payoff reel. There are innumerable designs for accomplishing the required performance. A schematic diagram of one arrangement in which a single motor supplies all the functions is shown in Fig. 8.10. The recording or playing condition is shown in Fig. 8.10A. The motor drive of the flywheel and capstan is accomplished by means of pressure idlers. There are two rotational speeds shown in Fig. 8.10A, which are obtained by either the fast or slow pressure idlers engaging the flywheel and motor shaft. The tape reel drive is provided by the pressure idlers engaging the flywheel and reel drive, and there is a brake on the payoff reel. The rewind condition is shown in Fig. 8.10B. Another set of pressure idlers connect the motor to the flywheel, and now the speed of the flywheel

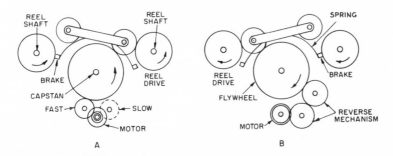

Fig. 8.10   The elements of the drive mechanism of a magnetic tape machine in which there is a single motor for driving the capstan, takeup reel, and payoff reel. A. Record and playback condition. B. Rewind condition.

is greater. A pressure idler engages the flywheel and the reel drive, and there is a brake on the other reel.

In other designs, belts are used instead of pressure idler wheels. The appropriate belts are activated for the various functions by means of idler pulleys. In some designs there are combinations of idlers, wheels, and belts.

Three magnetic heads are used in magnetic tape machines, namely, erasing, recording, and reproducing heads, as shown in Fig. 8.8. In reproducing, the erasing and recording heads are immobilized. In recording, any previous signal is removed by the erase head, the removal being accomplished by feeding a high frequency signal of high intensity to the erase head.

When the recording process is carried out with constant current in the head and the reproducing process is carried out with an amplifier in which the response is independent of the frequency, the overall response will be given by the characteristic of Fig. 8.6. Therefore, suitable compensation must be provided in order to obtain an overall uniform response frequency characteristic. The upper frequency limit is determined by the first dip, where the gap is equal to the wavelength. In recording, suitable high frequency accentuation is applied in the range above $d/\lambda = 0.3$ so that in reproduction no additional compensation will be required in this frequency range. In reproduction, the response is accentuated 6 dB per octave with decrease in frequency in the frequency range below $d/\lambda = 0.3$. In this way a uniform recording-reproduction characteristic is obtained. The accentuation of high frequency response in recording increases the signal-to-noise ratio. There is no overload problem in this frequency range due to accentuation of response in recording, because the amplitude of speech and music is lower in the high frequency range compared to the mid- and low frequency ranges. In the frequency range below $d/\lambda = 0.3$, the amplitude of flux ϕ will be constant for constant current in the recording head. Therefore, in order to obviate overloading of the tape in recording, the low frequency compensation must be supplied in the reproducing amplifier.

The standard tape speeds are as follows: 30, 15, $7\frac{1}{2}$, $3\frac{3}{4}$, and $1\frac{7}{8}$ inches per second. The higher speeds are used for high-quality recording. At the lower speeds the frequency bandwidth and signal-to-noise ratio are reduced. The standard tape width is $\frac{1}{4}$ inch. Single, double, quadruple, and octuple magnetic track recordings are used on the $\frac{1}{4}$-inch tape.

The upper frequency limit of reproduction will depend upon the air gap of the head and the tape speed. With heads in use today, and at tape speeds of 30 and 15 inches per second, an upper frequency limit of 25,000 hertz can easily be achieved. At a tape speed of $7\frac{1}{2}$ inches per second an upper-frequency limit of 20,000 hertz per second is possible in a well-designed system. The upper-frequency limit employing tape speeds of $3\frac{3}{4}$ and $1\frac{7}{8}$ inches is lower depending upon the air gap in the reproducing magnetic head.

The most common reel is made of plastic and is 7 inches in diameter. The 7-inch-diameter reel holds 1200 feet of 2.2 mil tape (1.5 mil base), 1800 feet of the 1.7 mil tape (1.0 mil base), 2400 feet of the 1.1 mil tape (0.8 mil base), and 3600 feet of the 0.7 mil tape (0.5 mil base). Tapes are sometimes rated according to base. There are also plastic reels 5 inches and 3 inches in diameter. Most professional machines use a metal reel 10.5 inches in diameter. There is also a metal reel 14 inches in diameter.

8.3 MONOPHONIC-SOUND MAGNETIC TAPE RECORDING AND REPRODUCING SYSTEMS

The elements of a complete monophonic-sound magnetic tape recording system are shown in Fig. 8.11. The sound is picked up by the microphone,

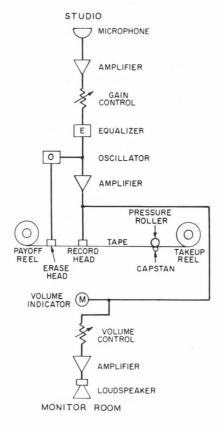

Fig. 8.11   The elements of a monophonic sound magnetic tape recording system.

amplified, and fed to the gain control. The output of the gain control is fed to an equalizer. Although the major compensation for the frequency response characteristic of Fig. 8.6 is carried out in reproduction, there is some accentuation of the very high frequency response provided in the equalizer. The outputs of the equalizer and the high frequency bias oscillator are fed to the power amplifier. The high frequency oscillator is of the transistor type operating at a frequency of 50 to 150 kilohertz. The power amplifier drives the magnetic recording head. The erase head actuated by the high frequency oscillator erases any previously recorded signal on the magnetic tape. A volume-indicating meter is used to monitor the level of the audio signal applied to the recording head. A loudspeaker is used to monitor the audio signal applied to the recording head. In some machines the amplifier and loudspeaker monitor are connected to the reproducer head so that the recorded signal is monitored. The magnetic-tape transport mechanism has been described in Section 8.2 and Figs. 8.8, 8.9, and 8.10.

Sectional views of the monophonic, single-channel, magnetic heads for recording single, double, and quadruple magnetic tracks on the tape are shown in Fig. 8.12. The magnetic head consists of a stack of laminations with an air gap in the portion where the magnetic head makes contact with the magnetic tape and another air gap in the rear portion of the head. Two coils surround the laminations. The electrical impedance of the magnetic head for recording, reproducing, and erase can be made appropriate for transistor amplifiers for the functions of reproducing, recording, and erasing.

Top views of the monophonic, single-channel, magnetic tracks for single,

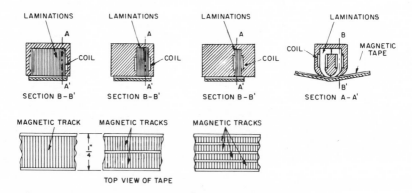

Fig. 8.12 Sectional views of monophonic, single channel, magnetic heads for reproducing single, double and quadruple magnetic tracks. The top view of the magnetic tape shows the single, double and quadruple magnetic tracks.

double, and quadruple magnetic tracks on the magnetic tape are shown in Fig. 8.12. In general, the single track is seldom used. Two and four magnetic tracks are the most common. In the two-magnetic-track configuration, the two magnetic tracks are recorded in opposite directions, a procedure which eliminates rewinding in the reproduction. In the four-track-configuration there are two tracks in one direction and two in the other direction, an arrangement which also eliminates rewinding in the reproduction.

The elements of a complete monophonic-sound magnetic tape reproducing system are shown in Fig. 8.13. The output of the magnetic head is amplified by the voltage amplifier. The output of the voltage amplifier is fed to an equalizer. The frequency response characteristic of a magnetic tape reproducing head without compensation is shown in Fig. 8.6. The frequency response characteristic of the equalizer for 15, $7\frac{1}{2}$, $3\frac{3}{4}$, and $1\frac{7}{8}$ inches per second tape speeds are shown in Fig. 8.14. Some additional compensation is usually introduced at the two extremes of the frequency range to provide a uniform overall frequency response characteristic. The output of the equalizer is fed to the volume control. The output of the volume control is fed to the power amplifier. The power amplifier drives the loudspeaker. The magnetic heads used in reproduction are essentially the same as those used in recording. The magnetic heads and magnetic tracks on the magnetic tape are shown in Fig. 8.12 and have been described above.

Transistor amplifiers, described in Chapter 5, are used exclusively in magnetic tape sound recording and reproducing systems.

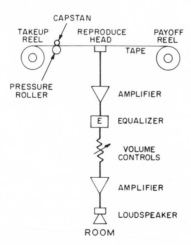

Fig. 8.13   The elements of a monophonic sound magnetic tape reproducing system.

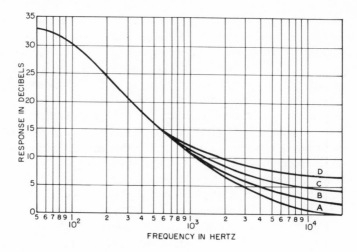

Fig. 8.14 The frequency response characteristic of the equalizer in the magnetic tape reproducing system for various tape speeds. A. Tape speed of 15 inches per second. B. Tape speed of $7\frac{1}{2}$ inches per second. C. Tape speed of $3\frac{3}{4}$ inches per second. D. Tape speed of $1\frac{7}{8}$ inches per second.

8.4 STEREOPHONIC-SOUND MAGNETIC TAPE RECORDING AND REPRODUCING SYSTEMS

The elements of a complete stereophonic-sound magnetic tape recording system are shown in Fig. 8.15. There are two identical channels of the type shown in Fig. 8.11 up to the magnetic tape recording head.

Sectional views of the stereophonic, two-channel, magnetic tape recording head are shown in Fig. 8.16. The two-channel magnetic recording head is in reality two heads of the type shown in Fig. 8.12. A top view of the magnetic tracks on the magnetic tape is also shown in Fig. 8.16. In this arrangement the tape must be rewound to be played.

Sectional views of a two-channel magnetic tape recording head in which there are four magnetic tracks on the magnetic tape are shown in Fig. 8.17. In the four-track arrangement, the two sets of magnetic tracks are recorded in opposite directions, a feature which eliminates rewinding in reproduction.

The elements of a complete stereophonic-sound magnetic tape reproducing system are shown in Fig. 8.18. Magnetic heads and tracks of the types shown in Figs. 8.16 and 8.17 are used in the reproduction, depending upon the magnetic track configuration.

The magnetic track configurations shown in Figs. 8.16 and 8.17 are for

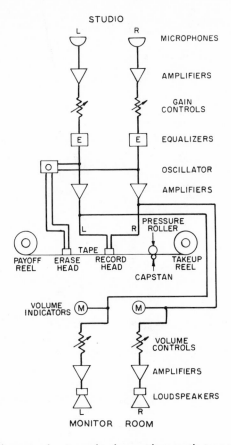

Fig. 8.15   The elements of a stereophonic sound magnetic tape recording system.

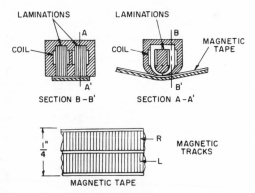

Fig. 8.16   Sectional views of a stereophonic, two channel magnetic head for reproducing double magnetic tracks. The top view of the magnetic tape shows the two magnetic tracks, *R* and *L*, representing the right and left channels.

magnetic tape on reels. There are also cartridges used in recording and reproducing magnetic tape. The magnetic heads and the magnetic tracks on the magnetic tape for stereophonic sound on magnetic tape in cartridges will be described in Section 8.7.

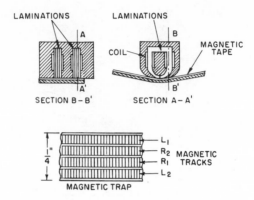

Fig. 8.17 Sectional views of a stereophonic, two channel magnetic head for reproducing a tape with quadruple magnetic tracks. The top view of the magnetic tape shows the four magnetic tracks. R_1 and L_1 are the right and left magnetic tracks in one direction, and R_2 and L_2 in the opposite direction.

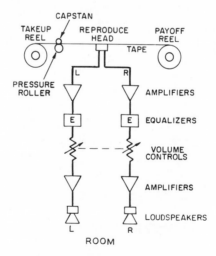

Fig. 8.18 The elements of a stereophonic sound magnetic tape reproducing system.

8.5 QUADRAPHONIC-SOUND MAGNETIC TAPE RECORDING AND REPRODUCING SYSTEMS

The elements of a complete quadraphonic-sound magnetic tape recording system are shown in Fig. 8.19. There are four identical channels of the type shown in Fig. 8.11 up to the recording head.

Sectional views of the magnetic head and the magnetic tracks on the magnetic tape for the quadraphonic-sound magnetic tape recording system are shown in Fig. 8.20.

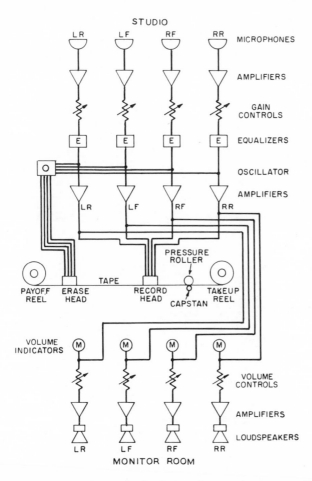

Fig. 8.19   The elements of a quadraphonic sound magnetic tape recording system.

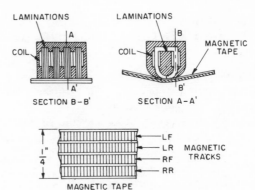

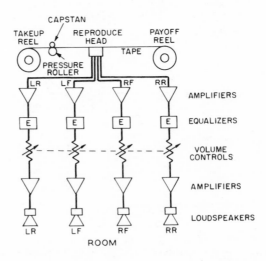

Fig. 8.20   Sectional views of a quadraphonic four channel magnetic head for reproducing quadruple magnetic tracks. The top view of the magnetic tape shows the four magnetic tracks, *LF*, *LR*, *RF* and *RR*, representing the channels left front, left rear, right front, and right rear, respectively.

The elements of a complete quadraphonic-sound magnetic tape reproducing system are shown in Fig. 8.21. The magnetic heads and the magnetic tracks on the magnetic tape are shown in Fig. 8.20. The track configuration shown in Fig. 8.20 is for quadraphonic sound on magnetic tape on reels. There is also a cartridge for reproducing quadraphonic sound on magnetic tape. The magnetic heads and the magnetic tracks on the magnetic tape for quadraphonic sound on magnetic tape in a cartridge will be described in Section 8.7.

Fig. 8.21   The elements of a quadraphonic sound magnetic tape reproducing system.

8.6 PROFESSIONAL MULTICHANNEL MAGNETIC TAPE SOUND-RECORDING

Multichannel magnetic tape sound-recording is employed in practically all professional master sound-recording for disk records and prerecorded magnetic tape. The first multichannel systems employed two or three channels. The three-channel recordings were made on magnetic tape $\frac{1}{2}$ inch in width. In the last few years eight-, sixteen-, and twenty-four-channel magnetic tape recorders have been used for master sound-recording. The use of the large number of channels makes it possible to record each musical instrument on a separate track. In the transfer to stereophonic and quadraphonic commercial records, all manner of modifications can be made to heighten the artistic impact. This subject will be considered in Chapter 15.

8.7 PRERECORDED SOUND ON MAGNETIC TAPE

Prerecorded monophonic, stereophonic, and quadraphonic sound on magnetic tape containing all manner of sound-program material have been commercialized on a wide scale. In the production of prerecorded sound on magnetic tapes, the master tape containing the sound program is reproduced by means of a master magnetic tape sound reproducer and fed to a bank of slave magnetic tape sound recorders. The master operates at a high tape speed of, say, 240 inches per second. The slaves operate at lower tape speeds of, say, 60 to 30 inches per second. In this way, a large number of tape reels or tape cartridges for tape speeds of $3\frac{3}{4}$ inches per second or tape cassettes for tape speeds of $1\frac{7}{8}$ inches per second are produced in one run of relatively short duration. The prerecorded sound on magnetic tape is packaged in the form of reels and cartridges as depicted in Fig. 8.22.

In Fig. 8.22A the magnetic tape is stored on plastic reels 3, 5, and 7 inches in diameter. The tape width is $\frac{1}{4}$ inch.

The prerecorded stereophonic sound on magnetic tape on reels is in the form of four magnetic tracks, with two magnetic tracks in one direction and two in the other direction. Sectional views of the two-channel head and the magnetic tracks on the magnetic tape are shown in Fig. 8.17. The stereophonic-sound magnetic tape reproducing system used for reproducing the two-channel magnetic tape on the reel is shown in Fig. 8.18. The tape speeds are $7\frac{1}{2}$ and $3\frac{3}{4}$ inches per second.

The prerecorded quadraphonic sound on magnetic tape on reels is of the four-track type. Sectional views of the four-channel head and the magnetic tracks on the tape are shown in Fig. 8.20. The quadraphonic-sound magnetic

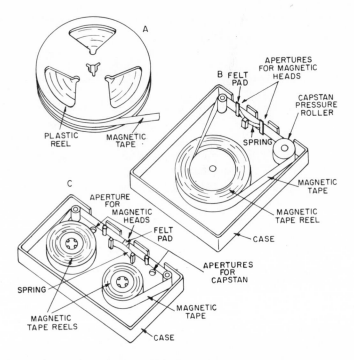

Fig. 8.22 Schematic perspective views of the means for the storage and packaging of magnetic tapes. A. Plastic reel. B. Continuous loop single reel cartridge. C. Two reel cartridge.

tape reproducing system used for reproducing the four-channel magnetic tape on the reel is shown in Fig. 8.21. The tape speed is $3\frac{3}{4}$ inches per second.

A (stereo-eight) continuous-loop magnetic tape cartridge is shown in Fig. 8.22B. There are open spaces in the cartridge case so that the capstan and magnetic heads can engage the magnetic tape. The rubber roller in the cartridge presses the tape against the capstan. In this way the magnetic tape is pulled from the center of the reel. The magnetic tape is fed back to the outside of the reel. In this system there is a continuous motion between adjacent layers of the magnetic tape with a resultant slippage between layers. The magnetic tape is lubricated to reduce friction between layers to satisfactory operable value.

Sectional views of the magnetic head and the magnetic tracks on the magnetic tape for the two-channel stereophonic sound on the continuous-loop magnetic tape cartridge are shown in Fig. 8.23. There are a total of eight tracks on the magnetic tape, which is $\frac{1}{4}$ inch in width. The two-channel head is shifted across the magnetic tape to reproduce the four sets of magnetic tracks. The stereophonic-sound magnetic tape reproducing system used for

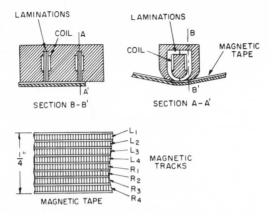

Fig. 8.23 Sectional views of the stereophonic magnetic head for reproducing the continuous loop cartridge of Fig. 8.22B. The top view of the magnetic tape shows the eight magnetic tracks. L_1 and R_1, L_2 and R_2, L_3 and R_3 and L_4 and R_4 represent four programs all in the same direction. The magnetic head is shifted laterally to reproduce each of the four programs.

reproducing the two-channel continuous-loop magnetic tape cartridge is essentially the same as the system shown in Fig. 8.18.

Sectional views of the magnetic head and the magnetic tracks on the magnetic tape for the four-channel quadraphonic sound on the continuous-loop magnetic tape cartridge are shown in Fig. 8.24. There are a total of

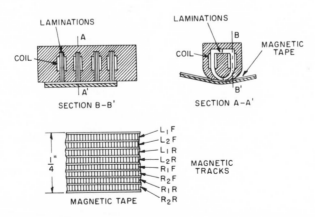

Fig. 8.24 Sectional views of the quadraphonic magnetic head for reproducing the continuous loop cartridge of Fig. 8.22B. The top view of the magnetic tape shows the eight magnetic tracks. L_1F, L_1R, R_1F and R_1R are the magnetic tracks for one program. The magnetic head is shifted laterally to reproduce the magnetic tracks L_2F, L_2R, R_2F and R_2R of the other program.

eight tracks on the magnetic tape, which is $\frac{1}{4}$ inch in width. The four-channel magnetic head is shifted across the magnetic tape to reproduce the two sets of magnetic tracks. The quadraphonic-sound magnetic tape reproducing system used for reproducing the four-channel continuous-loop magnetic tape cartridge is essentially the same as the system shown in Fig. 8.21.

A two-reel cassette magnetic tape cartridge is shown in Fig. 8.22C. The operation is essentially the same as that of the conventional reel system depicted in Fig. 8.8 except that reels are contained in a cartridge. There are open spaces in the cartridge case so that capstan drive and magnetic heads can engage the magnetic tape. This cartridge, termed the cassette, employs magnetic tape 0.15 inch in width. The tape speed is $1\frac{7}{8}$ inches per second.

Sectional views of the magnetic head and the magnetic tracks on the magnetic tape for the two-channel stereophonic sound on the cassette cartridge are shown in Fig. 8.25. There are four tracks on the magnetic tape. At the end of the program in one direction the cartridge is turned over and the other two tracks are reproduced. The two tracks in each direction are adjacent, in contrast to the other systems described in this chapter in which the tracks are staggered.

The stereophonic-sound magnetic tape reproducing system used for reproducing the two-channel magnetic tape on the cassette magnetic tape cartridge is shown in Fig. 8.18. The tape speed is $1\frac{7}{8}$ inches per second.

Sectional views of the magnetic head and the magnetic tracks on the magnetic tape for the four-channel quadraphonic sound on the cassette cartridge are shown in Fig. 8.26. There are eight tracks on the magnetic tape. At the end of the program in one direction the cartridge is turned over

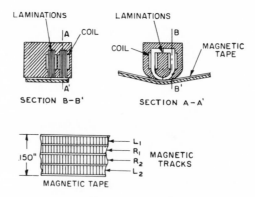

Fig. 8.25 Sectional views of the stereophonic magnetic head for reproducing the cassette cartridge of Fig. 8.22C. The top view of the magnetic tape shows the four magnetic tracks. R_1 and L_1 are for the magnetic tracks in one direction. The cartridge is turned over to reproduce the R_2 and L_2 magnetic tracks in the opposite direction.

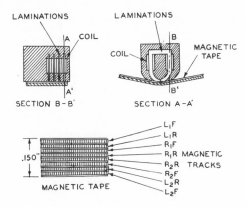

Fig. 8.26 Sectional views of the quadraphonic magnetic head for reproducing the cassette cartridge of Fig. 8.22C. The top view of the magnetic tape shows the eight magnetic tracks. L_1F, L_1R, R_1F and R_1R are the magnetic tracks in one direction. The cartridge is turned over to reproduce the L_2F, L_2R, R_2F and R_2R magnetic tracks in the opposite direction.

and the other four tracks are reproduced. The four tracks in each direction are adjacent.

The quadraphonic-sound magnetic tape reproducing system used for reproducing the four-channel magnetic tape on the cassette magnetic tape cartridge is essentially the same as the system shown in Fig. 8.21. The tape speed is $1\frac{7}{8}$ inches per second.

The cassette magnetic tape cartridge is also used for recording in both monophonic and stereophonic sound. In the monophonic-sound mode there are two magnetic tracks, one in each direction.

8.8 DISTORTION

There are many sources and types of distortion in magnetic tape sound-recording and reproducing systems. The most common are frequency discrimination, nonlinear amplitude, and frequency modulation. Frequency discrimination distortion due to a nonuniform frequency response was considered earlier in this text. Accordingly, nonlinear amplitude distortion and frequency modulation distortion will be considered in this section.

A. Nonlinear Amplitude Distortion

The magnetic medium of magnetic tape is of necessity nonlinear because it must possess retentivity in order to retain the magnetic signal applied to the magnetic tape. To overcome the nonlinearity depicted in Figs. 8.2 and 8.7, a high frequency ac bias is used as described in Section 8.2 and depicted in Fig. 8.7. The nonlinear amplitude distortion can be reduced to a low level by proper design and operation of the high frequency ac bias system. In a high-quality professional magnetic tape sound-recording and reproducing system, an overall nonlinear distortion of approximately 1 percent can be achieved.

Nonlinear amplitude distortion is also produced by "drop outs," that is, the absence of magnetic material coating in a small area of the tape. As a result there is a reduction in signal output when the magnetic reproducing head passes a void in the magnetic coating on the tape.

Print-through is the transfer of magnetism from a magnetized layer to adjacent layers of the tape. In this manner a small amount of the signal is magnetically recorded in adjacent layers. The net result is pre- and post-echo of the print through signal upon the main signal. Print-through is proportional to time that the tape has been wound on the reel without replaying and inversely proportional to the thickness of the base. Repeated rewinding of the tape at regular intervals reduces the effects of print-through by randomizing the print-through process. Certain types of processing when the magnetic coating is applied to the plastic base reduce the magnitude of the print-through. The level of the print-through is about 50 to 60 dB below the top level of the signal.

B. Frequency Modulation Distortion

Nonlinear distortion in the form of a frequency modulation of the signal is due to nonuniform motion of the magnetic tape past the magnetic recording head in recording, and nonuniform motion of the magnetic tape past the magnetic reproducing head in reproducing. This type of nonlinear distortion is termed flutter and wow. A frequency modulation of $\frac{1}{2}$ to 5 hertz per second is a wow, while above this frequency is a flutter.

The magnitude of the flutter or wow is given by:

$$f_K = \frac{df}{\sqrt{2} f_{AV}} \times 100 \qquad (8.3)$$

where f_K = flutter, in percent,
df = maximum deviation in frequency from the average frequency, in hertz, and
f_{AV} = average frequency, in hertz.

Obviously, flutter and wow can be reduced to a negligible subjective quantity by a high-quality tape transport mechanism. For high-quality sound reproduction the wow and flutter should be less than 0.4 percent.

8.9 NOISE

The magnetic tape coating on the plastic base consists of particles of magnetic oxide as depicted in Fig. 8.1. Noise in magnetic tape reproduction is produced in reproduction because of the random distribution of the magnetized particles on the tape. The noise power is inversely proportional to the tape speed and to the magnetic track width. For a tape speed of 15 inches per second and a track width of 0.1 inch, a signal-to-noise ratio of 60 dB may be obtained. The signal-to-noise ratio is proportional to the speed and the track width. When the speed is reduced by a factor of one-half, the signal-to-noise ratio is reduced by 3 dB. When the track width is reduced by one-half, the signal-to-noise ratio is reduced by 3 dB.

REFERENCES

Olson, Harry F. *Acoustical Engineering*, Van Nostrand Reinhold, New York, 1957.

Pear, C. B., Editor. *Magnetic Recording in Science and Industry*, Van Nostrand Reinhold, New York, 1967.

Stewart, W. E. *Magnetic Recording Techniques*, McGraw-Hill, New York, 1958.

Tremaine, Howard M. *Audio Cyclopedia*, Howard W. Sams, Indianapolis, 1969.

(See also more than two hundred articles on magnetic tape recording and reproducing technology and standards in the *Journal of the Audio Engineering Society*.)

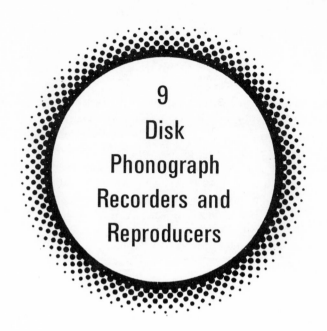

9
Disk
Phonograph
Recorders and
Reproducers

9.1 INTRODUCTION

A disk phonograph recorder consists of an electro-mechanical transducer equipped with a chisel-shaped stylus which cuts an audio frequency spiral groove in a plastic disk record. The motion of the stylus cuts modulations in the groove which correspond to the electrical signals applied to the transducer. A disk phonograph reproducer consists of a mechano-electric transducer equipped with a stylus operating in the groove of the plastic disk record; it thereby transforms the audio frequency modulations in the groove to the corresponding electrical output. The purpose of this chapter is to describe the processes and machines employed in disk phonograph recording and reproducing of audio signals. In view of the fact that nearly all the new disk records now being issued are stereophonic and practically all disk phonographs now being sold are stereophonic, the monophonic disk record and monophonic phonograph will not be considered in this book. Furthermore, monophonic disk phonograph records can be played on stereophonic phonographs, and stereophonic disk records can be played on monophonic phonographs. Quadraphonic sound reproduction by means of disk records, a recent development, will also be considered.

9.2 DISK PHONOGRAPH RECORDING PROCESS

A. Disk Phonograph Recording System

The elements of a complete stereophonic disk phonograph sound-recording system are shown in Fig. 9.1. In the case of direct recording upon the disk, the outputs of the microphones are amplified and fed to the gain controls and on to the rest of the system. However, in general, the recording of the disk record is not made directly from the live pickup. Instead the original recording is a two-channel stereophonic magnetic tape record. In this case, as

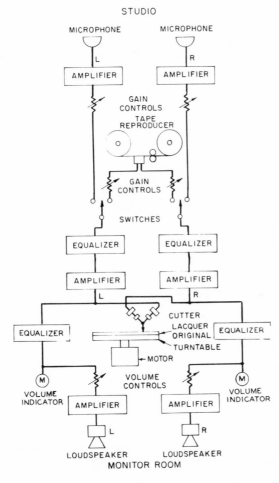

Fig. 9.1 The elements of a stereophonic disk phonograph recording system.

depicted in Fig. 9.1, the outputs of the magnetic tape reproducer are fed to equalizers. The outputs of the equalizers are amplified and fed to the cutter. The monitoring system consisting of the meter and loudspeaker provides means for checking the level and sounds of the audio signal applied to the cutter. The disk cutter cuts the modulated groove in the record. The rotation of the record is provided by the turntable and motor. The elements of the system will be described in detail in the sections which follow.

B. Disk Phonograph Recorder

An electrical disk phonograph recorder is a machine for transferring an electrical signal input to the corresponding modulated groove in a plastic disk record. An electrical disk phonograph recorder is depicted in Fig. 9.2. The lacquer disk used in recording the master record is placed upon the recording turntable. To provide uniform rotational motion the turntable is made very heavy, thereby providing a large rotational inertia. In general, the turntable is driven by a synchronous motor. A suitable rotational mechanical filter is placed between the motor and the turntable so that flutter and wow will be reduced to a negligible amount. As depicted in Fig. 9.2, the drive system is arranged so that all the standard record speeds can be cut. The lead screw drives the cutter in a radial direction so that a spiral groove is cut in the record. Different rotational speeds of the lead screw are provided so that 100 to 800 grooves per inch can be cut. In some recordings a variable

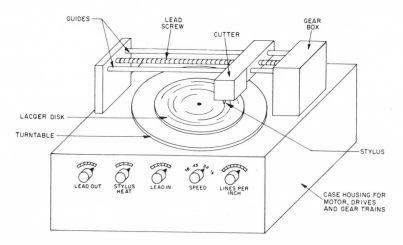

Fig. 9.2 Perspective schematic view of a stereophonic disk phonograph recorder. The particular design shown is for illustrative purposes in depicting the essential elements and does not necessarily correspond to any existing machine.

pitch is used. In this procedure the spacing between the grooves is made to correspond to the amplitude—small spacing for small amplitudes and large spacing for large amplitudes. Under these conditions the maximum amount of information can be recorded on the record. The material removed in the cutting process is in the form of a fine thread. The thread is pulled into the open end of a pipe near the cutting stylus, the other end of which is connected to a vacuum system.

C. Disk Phonograph Cutter

A sectional view, mechanical circuit, and electrical system, and the frequency response characteristic of a stereophonic disk phonograph cutter are shown in Fig. 9.3. The vibrating system in each channel is of the dynamic type with a driving coil, a sensing coil, and a suitable suspension system. The two vibrating systems are arranged at 90° with respect to each other. Therefore, each system operates independently of the other. Accordingly, the consideration of the vibrating system will be concerned with only one channel. The vibrating system is designed so that there is a single degree of freedom over the audio frequency range as shown by the mechanical circuit of Fig. 9.3. The velocity frequency response characteristic of Fig. 9.3 shows that the vibrating system

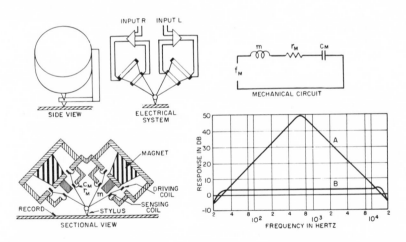

Fig. 9.3 Schematic view, side view, electrical system, mechanical circuit and velocity frequency response of a stereophonic disk phonograph recorder. In the mechanical circuit: f_M = driving force. m = mass of the voice coils and associated equipment. C_M = compliance of the suspensions and r_M = mechanical resistance of the suspension system. A. Velocity response of the stylus without feedback. B. Velocity response of the stylus with feedback.

behaves as a system of one degree of freedom from 30 to 16,000 hertz with the fundamental resonant frequency at 700 hertz. The output of the sensing coil is fed to the input of the amplifier. The output of the amplifier is fed to the driving coil in an out-of-phase relationship with the signal input. With the feedback in operation the velocity of the vibrating system is practically independent of the frequency over the frequency range of 30 to 16,000 hertz as shown in Fig. 9.3. The input to the amplifier can be compensated to provide the desired recording characteristic.

D. Cutting Stylus

The original recording is cut by means of a stylus attached to the cutter of Fig. 9.3. The cutting stylus consists of a sapphire, synthetic ruby, or other hard material shaped in the form of a pointed chisel as shown in Fig. 9.4. The stylus is heated in recording and thereby imparts a smooth side wall to the groove, which reduces the noise by a factor of 20 dB. To heat the stylus the shaft is wound with a few turns of fine wire and supplied by an electrical current from a low-voltage dc supply, as shown in Fig. 9.4D. (See also Fig. 9.2.)

E. Groove Undulations

In the two-channel recorder of Fig. 9.2, the two vibrating systems are arranged at right angles to each other. Therefore, the two channels are recorded on the two sides of the groove at right angles wholly independent of each other. The modes of vibration in a plane normal to the surface of the record and normal to the groove are shown in Fig. 9.5. The motion of the cutting stylus and the reproducing stylus is also depicted in Fig. 9.5.

F. Recording Characteristic

The frequency velocity characteristic of the modulation in the groove in the disk phonograph record provides the velocity at the tip of the stylus of the

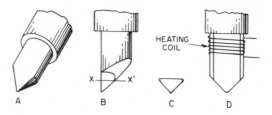

Fig. 9.4 Views of a cutting stylus. A. Perspective. B. Side. C. Section x-x'. D. Heating coil wound on the shaft of the stylus.

Fig. 9.5 Schematic views of groove undulations in a stereophonic disk phonograph record. Heavy lines indicate zero amplitude or unmodulated groove. Light lines indicate maximum limit of groove modulation. Arrows indicate direction of motion of recording stylus and reproducing stylus. A. Unmodulated groove. B. Modulation in right channel. C. Modulation in left channel. D. Lateral modulation, combination of B and C in phase. E. Vertical modulation, combination of B and C out of phase. F. Combination of equal vertical and lateral amplitudes, combination B and C with 90° difference in phase between B and C.

phonograph pickup as a function of the frequency. In the recording of commercial phonograph records high frequency compensation is employed in the frequency velocity characteristic. The Record Industry Association of America has established a standard recording frequency velocity characteristic as depicted in Fig. 9.6. In the reproduction of commercial disk phonograph records an inverse frequency response characteristic is employed in order to obtain a uniform over-all frequency response characteristic. Commercial standard frequency records exhibiting the RIAA frequency velocity characteristic are used in the development design and service of disk phonograph instruments. The frequency amplitude characteristic corresponding to the velocity is also shown in Fig. 9.6.

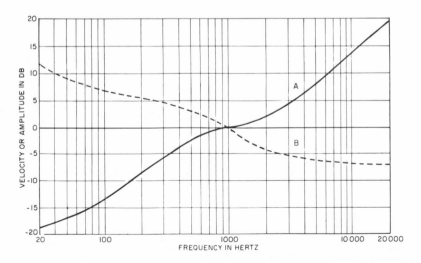

Fig. 9.6 Recording characteristics used in commercial disk phonograph records according to the Record Industry Association of America. A. Frequency velocity characteristic. B. Frequency amplitude characteristic.

9.3 DISK PHONOGRAPH RECORD MANUFACTURE

The processes in the record manufacturing plant for the mass production of records are depicted in Fig. 9.7. The original lacquer disk termed the "original" of Fig. 9.7A is metalized and then electroplated. The plating is separated from the lacquer and reinforced by backing with a solid metal plate. The assembly is termed the "master" (Fig. 9.7B). The master is electroplated. This plating is separated from the master and reinforced by backing with a solid metal plate. The assembly is termed the "mother" (Fig. 9.7C). Several mothers may be made from the master. The mother is electroplated. This plating is separated from the mother and reinforced by a solid metal plate. The assembly is termed the "stamper" (Fig. 9.7D). Several stampers may be made from each mother. One stamper containing a sound selection to be placed on one side of the final record is mounted in the upper jaw, and another stamper containing a sound selection to be placed on the other side of the record is placed in the lower jaw of a hydraulic press equipped with means for heating and cooling the stampers (Fig. 9.7E). A preform or biscuit of thermoplastic material such as a vinylite or other plastic compound is placed between the two stampers. The stampers are heated, and the jaws of the

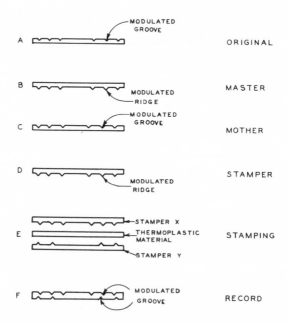

Fig. 9.7 The steps in the processes for the mass production of disk phonograph records from the lacquer original.

press are closed, thus pressing the two stampers against the thermoplastic material. When an impression of the stampers has been obtained in the thermoplastic material, the stampers are cooled, thus cooling and setting the plastic record. The jaws of the hydraulic press are opened, and the record is removed from the press. The modulated grooves in the record correspond to those in the original lacquer disk (Fig. 9.7F). The stamping procedure is repeated again and again until enough records are obtained. The process described constitutes the "mass-production system" for the production of phonograph records.

9.4 COMMERCIAL DISK PHONOGRAPH RECORDS

Commercial phonograph records as depicted in Fig. 9.8 are made in four speeds, namely, 78, 45, $33\frac{1}{3}$, and $16\frac{2}{3}$ revolutions per minute. The 78 rpm records are made in three diameters, termed 12-inch, 10-inch, and 7-inch. The nominal maximum playing times are 5, $3\frac{1}{2}$, and $2\frac{1}{2}$ minutes, respectively. The $33\frac{1}{3}$ rpm records are made in three diameters, termed 12-inch, 10-inch, and 7-inch. The nominal maximum playing times are 25, 17, and 8 minutes respectively. The 45 rpm record is made in a diameter of 7 inches. The

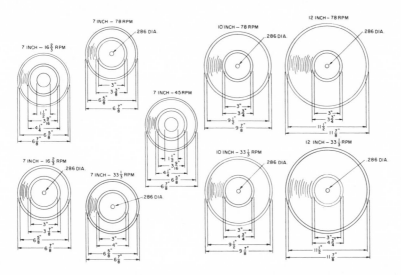

Fig. 9.8 Typical dimensions of the most common commercial-type disk phonograph records. The dimensions are the diameter of the outside of the record, the diameter of the outside and inside groove, the diameter of the label and the diameter of the center hole.

nominal maximum playing time is 8 minutes. The $16\frac{2}{3}$ rpm records are made in a diameter of 7 inches. The nominal maximum playing time of the records with the large center hole is 30 minutes. The nominal maximum playing time for the small-hole records is 45 minutes for music and 60 minutes for speech. The diameter of the outside, the diameters of the first and last program grooves, the label diameters, and the diameter of the center hole of the different records are shown in Fig. 9.8. It may be mentioned in passing that the specifications of Fig. 9.8 are given as representative and do not include all the variations.

The dimensions of the coarse groove, the fine groove, the stereo groove, and the ultra-fine groove and the dimensions of the corresponding styli are shown in Fig. 9.9. The coarse groove is used in 78 rpm records. The fine groove is used in the monophonic 45 and $33\frac{1}{3}$ rpm records. The stereo groove is used in the stereophonic 45 and $33\frac{1}{3}$ rpm records. The ultra-fine groove is used in the $16\frac{2}{3}$ rpm record.

The stylus radius as shown in Fig. 9.9 is 0.75 mil for stereophonic reproduction. A stylus with an elliptical cross section for stereophonic reproduction has been introduced as depicted in the cross section of Fig. 9.9. The radius in contact with the record is 0.25 mil. Employing a smaller radius reduces the tracing nonlinear distortion. (See Section 9.6C.) However, since the area of the stylus in contact with the record surface is less for the elliptical stylus, the indications are that the record wear will be greater for the same stylus force. The stylus material is sapphire or diamond.

The maximum nominal grooves per inch for the different size grooves are as follows: coarse groove, 125; fine and stereo groove, 275; and ultra-fine groove, 550.

The maximum amplitudes, in inches, in the frequency range 200 to 2000 hertz for the different size grooves are as follows: coarse groove, 0.004–0.005 inch, fine and stereo groove, 0.0015–0.002 inch; and ultra-fine groove, 0.0007–0.001 inch.

All of the records depicted in Fig. 9.8 are commercially available. However, the modern issues of new programs except for a very small fraction are produced on the 12-inch $33\frac{1}{3}$ rpm, the 7-inch $33\frac{1}{3}$ rpm, and the 7-inch 45 rpm. Furthermore, present-day records are nearly all recorded in stereophonic sound.

9.5 DISK PHONOGRAPH REPRODUCING PROCESS

A. Disk Phonograph Reproducing System

The elements of a complete stereophonic disk phonograph sound-reproducing system are shown in Fig. 9.10. The turntable is rotated at a constant speed by means of the motor. The stylus follows the undulations in the groove and transfers the vibrations of the stylus to the pickup transducers. The outputs of the pickup are amplified and fed to ganged volume controls. The volume control is followed by an equalizer. The output of the equalizers are amplified and fed to the loudspeakers. The elements of the system will be described in the sections which follow.

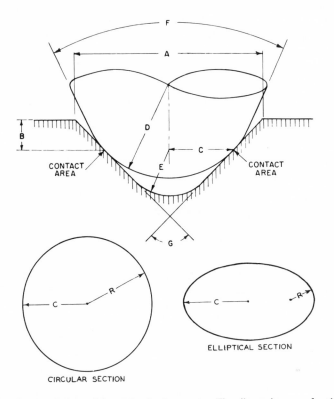

Fig. 9.9   Sectional view of the stylus in the groove. The dimensions are for the coarse groove, fine groove, stereo groove, and ultra-fine groove as follows:

Dimension	Coarse	Fine	Stereo	Ultra-Fine
A	.006	.0027	.0025	.001
B	.0008	.0006	.0005	.0004
C	.0019	.0007	.0005	.00017
D	.0027	.001	.00075	.00025
E	.001	.00027	.00027	.00015
F	45°	45°	45°	45°
G	90°	90°	90°	90°

The sectional views depict the sectional views of styli with circular and elliptical cross sections. The radius of the stylus contact with the groove is *R*.

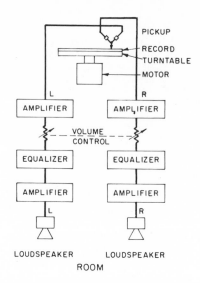

Fig. 9.10 The elements of a stereophonic disk phonograph reproducing system.

B. Disk Phonograph Record Player and Changer

An electrical disk phonograph record player and changer are machines for reproducing a disk phonograph record in which the electrical signal output corresponds to the modulations in the groove of the disk record. A disk phonograph record player and a changer are depicted in Fig. 9.11. The turntable is driven by a suitable speed-reduction system in the form of a belt in Fig. 9.12A, or in the form of a rubber idler in Fig. 9.12B, or in a combination of the belt and rubber idler. The drive system is designed to provide

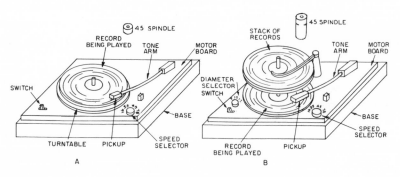

Fig. 9.11 A. Perspective schematic view of disk phonograph record player. B. Perspective schematic view of a disk phonograph record changer.

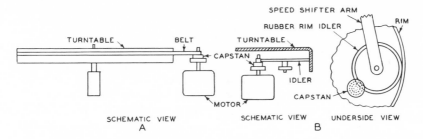

Fig. 9.12 Schematic views depicting the elements of variable turntable drives. A. Belt drive system. B. Idler drive system.

uniform rotational speed and isolation of the motor from the turntable to reduce rumble noise. In general, there are four speeds of rotation, namely, $16\frac{2}{3}$, $33\frac{1}{3}$, 45, and 78 rpm. The speed-changing mechanism supplies means for selecting the desired speed. The disk phonograph record player shown in Fig. 9.11A is used to play one record at a time. The record changing is carried out manually. The disk phonograph record changer shown in Fig. 9.11B plays a large group of records in sequence. When a record has been played, the tone arm enters the oscillating silent groove. The oscillation of the tone arm actuates the record change mechanism, which moves the pickup arm out beyond the record. The next record is tripped by the trigger on the shaft and falls to the turntable. The tone arm moves over to the lead in the groove in the record, and the record is played. The tone arm is pivoted at the rear end and carries the pickup at the front end. The design of the tone arm will be described in the next section.

C. Tone Arm

Top and side views of a tone arm are shown in Fig. 9.13. The vertical force which the stylus exerts upon the record is determined by the counterweight and the spring. In some tone arms the spring is omitted and the required stylus force is obtained by the position of the counterweight. In other tone arms the counterweight is fixed and the required stylus force is obtained by means of the tension in the spring. In some designs, the spring is in the form of a spiral around the pivot.

A source of nonlinear distortion due to a deviation in tracking, is commonly termed tracking error. The angle between the vertical plane containing the vibration axis of the pickup (for modulation D of Fig. 9.5) and the vertical plane containing the tangent to the record groove is a measure of the tracking error. If the vibration axis of the pickup passes through the tone arm pivot, the tracking error can be zero for only one point on the record. The tracking

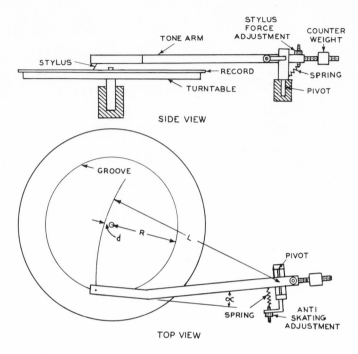

Fig. 9.13 Side and top views of a tone arm showing the counter weight and spring for providing the desired stylus force and the spring for countering the side thrust force. The geometry depicts the means for reducing the tracking error.

error can be reduced if the vibration axis of the pickup is set at an appropriate angle with respect to the line connecting the stylus point and the tone arm pivot together with a suitable overhang distance between the stylus and the record axis. (See Fig. 9.13.)

The amount of overhang d, in inches, is given by:

$$d = \frac{R_i^2}{L\{\frac{1}{4}[1 + (R_i/R_o)]^2 + (R_i/R_o)\}} \tag{9.1}$$

where L = length of the tone arm from the pivot to the stylus, in inches,
           R_o = radius of the start groove of the recording, in inches, and
           R_i = radius of the end groove of the recording, in inches.

The appropriate angle α, in degrees, between the vertical plane containing vibration axis of the pickup and the line joining the stylus and the tone arm pivot is given by:

$$\alpha = \frac{57.3[1 + (R_i/R_o)]R_i}{L\{\frac{1}{4}[1 + (R_i/R_o)]^2 + (R_i/R_o)\}} \tag{9.2}$$

With the application of Equations (9.1) and (9.2) in the design of a tone arm, the tracking error can be reduced to ± 5 percent.

There is also a tone arm in which the elements are articulated so that the pickup stylus is kept perpendicularly tangent to the groove throughout the record. In this system, a vertically pivoted pickup case is coupled by means of a thin rod to a point near the tone arm vertical pivot and thereby maintains a tangental pickup-groove configuration.

There is another factor involved in the tone arm, namely, side thrust due to the geometry of the record, pickup, and tone arm. The side thrust introduces a mechanical bias upon the pickup and may lead to nonlinear distortion. To counter the side thrust, a mechanical bias in the form of an outward force is applied to the tone arm as shown in Fig. 9.13. When this counter force is correct, the pickup can be placed upon a plain disk and the tone arm will remain at almost any part of the smooth unmodulated record without skating sidewise. A spring is depicted in Fig. 9.13 for providing the sidewise counter thrust. This spring may also be in the form of a spiral around the vertical pivot. Other means for providing the counter thrust may be opposed permanent magnets or a weight, string, and pulley arrangement.

D. Phonograph Pickup

A phonograph pickup consists of a mechano-electrical transducer equipped with a stylus operating in the groove of the plastic disk record, by which the modulations in the groove are converted to the corresponding electrical output. The most common phonograph pickups employ dynamic, magnetic, and ceramic transducers. Stereophonic phonograph pickups employing these transducers will be described in the sections which follow.

1. Dynamic Phonograph Pickup. Schematic views, mechanical network, electrical circuit, and the frequency response characteristic of a stereophonic dynamic phonograph pickup are shown in Fig. 9.14. The stereophonic dynamic phonograph pickup employs a very simple vibrating system and provides excellent velocity response over the audio frequency range as depicted in Fig. 9.14. The cross-talk between the two channels is relatively large and quite adequate. The electrical generator may be considered to be the open-circuit voltage in series with the electrical impedance of the voice coil. The voice coil is practically a constant electrical resistance over the audio frequency range. The electrical impedance of the voice coils in the pickup are relatively low, being of the order of a few ohms. Therefore, the electrical voltage output is also very low. A transformer is used to step up the impedance to that suitable for the input to a transistor amplifier. The use of a transformer also steps up the voltage applied to the input of the

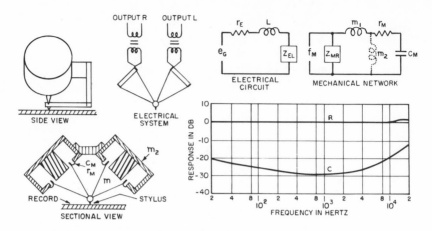

Fig. 9.14 Sectional view, side view, electrical system, mechanical network, electrical circuit and frequency response characteristics of a dynamic disk phonograph pickup. In the mechanical network: f_M = force generated by the velocity generator. z_{MR} = mechanical impedance of the record. m_1 = mass of the voice coil, stylus and associated connection. C_M = compliance of the suspension. r_M = mechanical resistance of the suspension. m_2 = mass of the pickup and tone arm. In the electrical circuit: e_G = open circuit voltage developed in the voice coil. L = inductance of the voice coil. r_E = resistance of the voice coil. z_{EL} = electrical impedance of the load. In the graph: R = output frequency response for a constant velocity record. C = cross talk frequency response characteristic.

amplifier. The voltage output of the dynamic pickup is proportional to the velocity of the stylus. Therefore, to reproduce the commercial disk phonograph records with the pickup of Fig. 9.14 without frequency discrimination or accentuation, the equalizer in Fig. 9.10 must exhibit the inverse of the frequency velocity characteristic of Fig. 9.6A.

2. Magnetic Phonograph Pickup. A schematic perspective view, mechanical network, electrical circuit, and frequency response characteristic of a stereophonic magnetic phonograph pickup are shown in Fig. 9.15. The vibrating system of the stereophonic magnetic phonograph pickup is a very simple one. The motion of the magnet changes the magnetic flux through the coils and thereby induces a voltage in the coils. There are many other designs of the variable magnetic reluctance system but the operation is essentially the same as in Fig. 9.15. The velocity frequency response characteristic is smooth over the audio frequency range as shown in Fig. 9.15. The cross talk between the two channels is relatively large and quite adequate. The electrical generator may be considered to be the open-circuit voltage in series with the electrical impedance of the coils. Since the electrical impedance is primarily inductive, the electrical impedance is proportional to the fre-

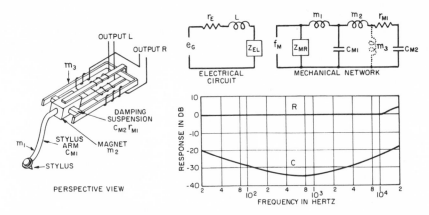

Fig. 9.15 Perspective view, mechanical network, electrical circuit and frequency response characteristics of a magnetic disk phonograph pickup. In the mechanical network: f_M = force generated by the velocity generator. z_{MR} = mechanical impedance of the record. m_1 = mass of the stylus and holder. C_{M1} = compliance of the stylus and tone arm. m_2 = mass of the magnet. C_{M2} = compliance of the suspension. r_{M1} = mechanical resistance of the damped suspension. m_3 = mass of the pickup and tone arm. In the electrical circuit: e_G = open circuit voltage developed in the coils. L = inductance of the coils. r_E = electrical resistance of the coils. z_{EL} = electrical impedance of the load. In the graph: R = output frequency response characteristic for constant velocity record. C = cross talk frequency response characteristic.

quency. The coils of the magnetic pickup can be wound to an electrical impedance suitable for the input of an amplifier. The voltage output of the magnetic pickup is proportional to the velocity of the stylus. Therefore, to reproduce the commercial disk phonograph records with the pickup of Fig. 9.15 without frequency discrimination or accentuation, the equalizer in Fig. 9.10 must exhibit the inverse of the frequency velocity characteristic of Fig. 9.6A.

3. Ceramic Phonograph Pickup. A schematic perspective view, mechanical network, electrical circuit, and frequency response characteristic of a stereophonic ceramic phonograph pickup are shown in Fig. 9.16. The vibrating system of the stereophonic ceramic phonograph pickup is somewhat more complex than the dynamic and magnetic stereophonic phonograph pickups. Therefore, considerable care must be taken in the design to provide a uniform frequency response characteristic. Very small ceramic elements (see Section 4.3C) are employed, the resonant frequency of which may be placed in the upper part of the audio frequency range. The cross talk is greater than for the dynamic or magnetic pickups but is quite adequate. The electrical generator may be considered to be an open-circuit voltage in series with the electrical capacitance of the ceramic element. The voltage output is proportional to the

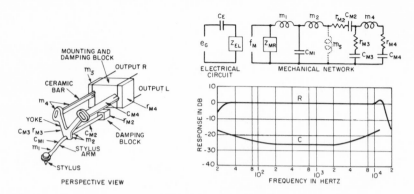

Fig. 9.16 Perspective view, mechanical network, electrical circuit and frequency response characteristic of a ceramic disk phonograph pickup. In the mechanical network: f_M = force generated by the velocity generator. z_{MR} = mechanical impedance of the record. m_1 = mass of the stylus and stylus holder. C_{M1} = compliance of the stylus arm. m_2 = mass of the yoke. C_{M3} and r_{M3} = compliance and mechanical resistance of the yoke. C_{M2} = compliance of the rear part of the stylus arm. r_{M2} = mechanical resistance of the damping block. m_4, r_{M4} and C_{M4} = mass, mechanical resistance and compliance of ceramic element and mounting. m_5 = mass of the pickup and tone arm. In the electrical circuit: e_G = open circuit voltage developed by the ceramic element. C_E = electrical capacitance of the ceramic element. r_{EL} = electrical impedance of the load. In the graph: R = output response frequency characteristic for constant amplitude record. C = cross talk frequency response characteristic.

displacement of the ceramic element. Therefore, the response as depicted in Fig. 9.16 is a frequency amplitude characteristic. Therefore, to reproduce the commercial disk phonograph records with the pickup of Fig. 9.16 without frequency discrimination or accentuation, the equalizer of Fig. 9.10 must exhibit the inverse of the frequency amplitude characteristic of Fig. 9.6B.

E. Compliance of Phonograph Pickups

A consideration of the stereophonic phonograph pickups in the preceding sections shows that the mechanical impedance at the stylus in the low frequency range is due to the compliance of the pickup. The compliance determines the force applied to the record in the frequency region where the pickup is stiffness-controlled. The force is the ratio of the displacement to the compliance. The compliance of a stereophonic phonograph pickup is of the order of more than 1×10^{-6} centimeters per dyne.

F. Tone Arm Resonance

Tone arm resonance occurs in the low frequency range when the mechanical reactance due to the effective mass of the pickup and tone arm is equal to the mechanical reactance of the compliance of the pickup. The response at the resonant frequency is determined by the damping in the pickup and tone arm pivot. The tone arm resonant frequency determines the low frequency range because the response falls off rapidly below the tone arm resonance.

G. Stylus Tracking Force

The stylus should track the groove modulation in a proper manner with a relatively small tracking force applied by the stylus to the record. If the force is large, the record wear will be accelerated. If the force is small, the stylus will not track because the stylus actually leaves the groove at small intervals with a resultant high nonlinear distortion. The design of a high-quality pickup is carried out so that the stylus will track the modulation of practically all commercial records with a tracking force of one gram.

H. Translation Loss

The finite mechanical impedance of the record leads to a reduction in high frequency response and a high frequency cutoff. The translation loss is due to the shunting of the mechanical impedance, z_{MR}, of the record in Figs. 9.14, 9.15, and 9.16. At the very high frequencies the mechanical impedance, z_{MR}, becomes smaller than the mechanical impedance of the phonograph pickup at the stylus. As a result, there is a reduction in high frequency response. For commercial records of vinylite and for normal operation of high-quality phonograph pickups, the translation loss is negligible in the audio frequency range below 15,000 hertz.

I. Scanning Loss

Scanning loss occurs when the wavelength in the groove becomes comparable to the dimension of the contact surface of stylus with groove surface. The scanning loss in commercial records of vinylite for normal operation of high-quality phonograph pickups is negligible in the audio frequency range below 15,000 hertz.

9.6 NONLINEAR DISTORTION

The recording and reproducing of a phonograph record is a complicated process and there are many sources of nonlinear distortion. The major sources of nonlinear distortion in a disk phonograph sound-recording and reproducing system are a finite mechanical impedance of the record material, pickup-tone arm tracking, stylus tracing, vertical tracking, and nonuniform speed of the groove past the stylus.

A. Finite Mechanical Impedance of the Record Material

The record does not present an infinite mechanical impedance to the stylus. As a result, the vibrating system of the pickup is shunted by the mechanical impedance, z_{MR}, of the record, as depicted in the mechanical networks of pickups in Figs. 9.14, 9.15, and 9.16. Nonlinear distortion will occur if the record is a variable element. If the force which the stylus presents to the record is of such magnitude that it exceeds the yield point of the record material, the mechanical impedance of the record will not be a constant. The result is production of nonlinear distortion. Furthermore, if the force exceeds the yield point by a considerable amount, the record may be permanently damaged.

B. Tracking Nonlinear Distortion

The angle between the plane which is normal to the vibrating axis (for modulation D of Fig. 9.5) of the pickup stylus and the plane normal to the groove is termed the tracking angle error. The means for reducing the nonlinear distortion due to the tracking error was discussed in Section 9.5C.

C. Tracing Nonlinear Distortion

The master stereophonic disk termed the "original" is cut with a chisel-type stylus, and a disk record replica is reproduced with a ball-tipped stylus. In the stereophonic disk phonograph system, the vibration is of the vertical type. Therefore, there is a discrepancy between the motion of the cutting stylus and the reproducing stylus which becomes more pronounced for the shorter wavelength of the modulation in the groove. This type of nonlinear distortion is termed tracing nonlinear distortion.

The recording and reproducing process in one channel of a stereophonic disk phonograph system is shown in Fig. 9.17. The input signal e_1 to the recording system is a sine wave. The groove g_1 cut in the lacquer disk original record is also a sine wave, as is the groove g_2 in the disk record replica of the master original. However, the path of the ball-tipped stylus is not a sine wave, but is distorted in a manner characteristic of a vertical recording system. The electrical output e_2 is unsymmetrical, which means the major nonlinear distortion component is due to the second harmonic.

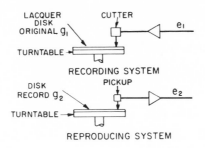

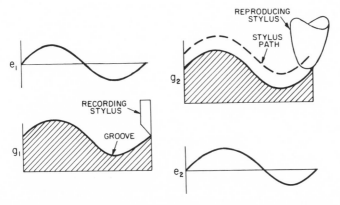

Fig. 9.17   The recording and reproducing process in one channel of a disk phonograph system. e_1 = voltage input to the recording system. g_1 = groove cut in the lacquer disk original. g_2 = groove in the disk record. e_2 = voltage output of the reproducing system.

The nonlinear distortion illustrated in Fig. 9.17 can be reduced by the introduction of complementary nonlinear distortion, as depicted in Fig. 9.18. The input signal e_1 is again a sine wave. However, the dynamic styli correlator introduces nonlinear distortion which is added to e_1. Therefore, the output of the adder and the input to the cutter e_2 is not a sine wave. The groove g_1 cut in the lacquer disk original record corresponds to e_2 and is not a sine wave. The groove g_2 in the disk record replica of the lacquer disk original is also not a sine wave. The groove g_2 in the record is distorted by the dynamic styli correlator so that the path of the reproducing stylus is a sine wave. The electrical output e_3 is a sine wave. The mechanism of Fig. 9.18 indicates that the introduction of complementary nonlinear distortion in the recording process will reduce nonlinear distortion in the reproducing process. The nonlinear distortion is reduced by a factor of more than four-to-one in actual practice.

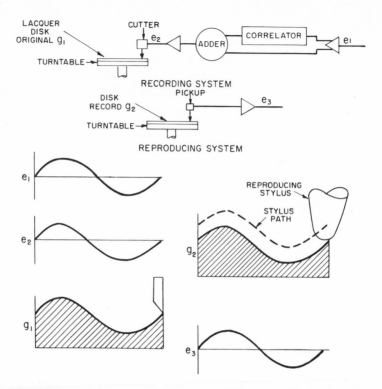

Fig. 9.18 The recording and reproducing process in one channel of a disk phonograph system employing a dynamic styli correlator to reduce the nonlinear distortion. e_1 = voltage input to the recording system. e_2 = voltage applied to the cutter. g_1 = groove cut in the lacquer disk original. g_2 = groove in the disk record. e_3 = voltage output of the reproducing system.

D. Vertical-Tracking-Angle Nonlinear Distortion

A disparity between the effective vertical angle in the recording cutter cutting the modulated groove in the record and the vertical tracking angle of the pickup will introduce harmonic nonlinear distortion and intermodulation nonlinear distortion in the output of the phonograph pickup. The standard vertical angle in stereophonic disk records is 15°, that is, 15° off the vertical. Therefore, the angle between the axis of rotation of the vibrating system of the pickup and the plane of the record should be 15° in order to eliminate nonlinear distortion.

E. Groove-Speed-Variation Nonlinear Distortion

Nonlinear distortion is produced when there is a lack of correspondence in the groove speed in the recording and reproducing of a disk phonograph record. This type of nonlinear distortion leads to frequency modulation of the reproduced signal and is termed a wow. Wow may be produced due to a nonuniform rotational speed of the record in recording and/or reproducing. Wow may also be produced by misplacement of the center hole or groove configuration distortion during the manufacture of the disk phonograph record. The magnitude of the wow in percent is given by Equation (8.3) in Section 8.8B. In general, the main source of wows in a disk phonograph sound-reproducing system is a nonuniform rotational velocity of the reproducing turntable.

9.7 NOISE

There are two main sources of noise in disk phonograph sound reproduction as follows: specks of dirt or foreign material lodged in the groove or a pocket in the groove, both of which produce a "tick" in the reproduced sound and a random nonuniformity of the record walls which leads to a random noise or hiss. Careful processing and handling of the disk phonograph record will eliminate "ticks" in reproduction. The random signal-to-noise ratio in commercial disk phonograph reproduction that is due to the disk is of the order of 65 to 70 dB. Therefore, noise in stereophonic disk records is not a problem for the reproduction of sound in the home by means of disk phonograph records.

A low frequency noise, termed rumble, is produced in disk phonograph reproduction by the turntable, pickup, and preamplifier. In general, rumble noise is generated by the motor turntable drive. The rumble noise should be at least 40 dB below the maximum signal level.

9.8 COMPATIBILITY OF MONOPHONIC AND STEREOPHONIC DISK RECORDS AND PHONOGRAPHS

The monophonic disk phonograph record is a single-channel record in which the motion of the cutter in recording and the stylus in reproducing is normal to the groove and parallel to the plane of the disk. (See Fig. 9.5D.) The monophonic system has not been considered in this chapter for the reasons given in Section 9.1. Monophonic and stereophonic disk records and phonographs are compatible as follows:

When a single-channel monophonic record is reproduced by a single-channel monophonic phonograph system, the single-channel output is reproduced by the loudspeaker.

When a single-channel monophonic record is reproduced by the two-channel stereophonic phonograph system, the single-channel output is reproduced by the two loudspeakers.

When a two-channel stereophonic record is reproduced by a single-channel monophonic phonograph reproducing system, the sound reproduced by the single loudspeaker is the sum of the two sound programs originally recorded on the two channels of the stereophonic recording system.

When the two-channel stereophonic record is reproduced by a two-channel stereophonic reproducing system, the stereophonic sound program is reproduced on the separate channels corresponding to the recording channels, and the sound which emanates from the two loudspeakers corresponds to the sound picked up by the respective microphones.

9.9 QUADRAPHONIC DISK RECORDING AND REPRODUCING SYSTEMS

The advent of quadraphonic sound reproduction by means of magnetic tape systems has led to the development of quadraphonic sound reproduction by means of the disk record. There are two approaches to quadraphonic sound reproduction by means of the disk record, namely, the encoding from four channels to recording on a two-channel disk and reproducing from a two-channel disk and decoding to four channels, and the recording and reproducing of four discrete channels by means of the disk. These two developments will be described in the text which follows:

A. Quadraphonic Disk System Employing Coding and Matrixing

There are many different ways of matrixing and encoding from four channels to two channels and decoding and matrixing from two channels to four channels. The generic recording and reproducing systems employing coding and matrixing in these processes are depicted in Fig. 9.19. The four inputs LF, LR, RF, and RR, representing the left front, left rear, right front, and right rear respectively, are matrixed and encoded to the left, L, and right, R, channels of the conventional two-channel stereophonic recording system. Each of the two channels R and L contains a mixture of the four channels in such a manner that the four channels can be reconstituted in reproduction by means of the decoder and matrix. In all of the coding and matrixing systems, the separation between LF, LR, RR, and RF is of the order of 6 dB more or less, depending upon the program. As indicated in Section 6.7, the maximum information that can be transmitted by two channels is a fixed quantity. In the system of Fig. 9.19 the information of two channels is allocated to four channels. Therefore, the amount of information in each

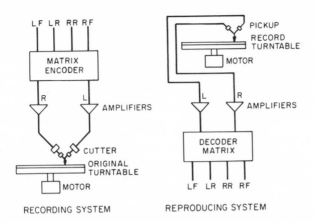

RECORDING SYSTEM   REPRODUCING SYSTEM

Fig. 9.19 The elements of a quadraphonic disk phonograph recording system employing matrixing and encoding from four channels to two channels and decoding and matrixing to four channels from two channels in reproducing.

channel is less than would be the case for four discrete channels. The performance, as exemplified by the auditory perspective and other spatial effects, of a two-channel-to-four-channel coded and matrixed system is inferior to that of four discrete channels.

B. Quadraphonic Disk System Employing Four Discrete Channels

A quadraphonic disk system in which two channels are recorded in the conventional manner and the other two channels are recorded by modulation means in the region above 15,000 hertz is shown in Fig. 9.20. The left front, *LF*, plus left rear, *LR*, signals are recorded in the left groove wall, and the right front, *RF*, plus right rear, *RR*, signals are recorded in the right groove wall in the conventional stereophonic mode and frequency range. The *LF* minus *LR* signals are recorded on a modulated carrier in the left groove wall, and *RF* minus *RR* signals are recorded on a modulated carrier in the right groove wall. In reproduction the *RF* plus *RR* are separated by means of a low pass filter. *RF* minus *RR* are separated and detected by means of high pass filter and detector. The two signals are fed to a matrix. The output is the discrete *RF* and *RR* signals. In the same way the discrete *LF* and *LR* signals are obtained from the left groove. The system is also compatible with a two-channel stereophonic system. For example, if the four-channel record is played on a conventional two-channel stereophonic system, the *LF* and *LR* will be reproduced on the left channel, and *RF* and *RR* will be reproduced on the right channel. If a conventional two-channel stereophonic

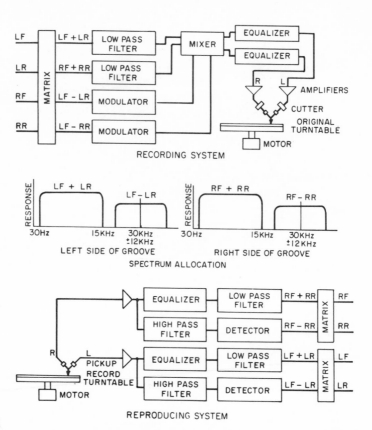

Fig. 9.20 The elements of a quadraphonic disk phonograph recording system in which two channels are recorded in the conventional manner and the other two channels are recorded by modulation means. The spectrum allocation depicts the channels and the frequency ranges. In reproducing, the two channels are reproduced in the conventional manner, and the other two channels are detected and reproduced.

record is played on the four-channel system the right channel will reproduce the same *R* signal on the *RF* and *RR* loudspeakers, and the left channel will reproduce the same *L* signal on the *LF* and *LR* loudspeakers.

REFERENCES

Bastians, C. R. "Factors Affecting the Stylus Groove Relationship in Phonograph Playback Systems," *Journal of the Audio Engineering Society*, Vol. 15, No. 4, p. 389, 1967.

Bauer, B. B. "Tracking Angle in Phonograph Pickups," *Electronics*, Vol. 18, No. 3, p. 110, 1945.

Fox, E. C. and Woodward, J. G. "Tracing Distortion—Its Causes and Correction in Stereodisk Recording Systems," *Journal of the Audio Engineering Society*, Vol. 11, No. 4, p. 294, 1963.

Frayne, John G. and Davis, R. R "Recent Developments in Stereo Disk Recording," *Journal of the Audio Engineering Society*, Vol. 7, No. 4, p. 147, 1959.

Inoue, T., Takahashi, N., and Owaki, I. "A Discreet Four-Channel Disk and Its Reproducing System," *Journal of the Audio Engineering Society*, Vol. 18, No. 1, p. 696, 1970.

Nelson, C. S. and Stafford, J. W. "The Westrex 3D Stereodisk System," *Journal of the Audio Engineering Society*, Vol. 12, No. 3, 1964.

Olson, Harry F. "The RCA Victor DYNAGROOVE System," *Journal of the Audio Engineering Society*, Vol. 12, No. 2, p. 98, 1964.

Olson, Harry F. *Solutions of Engineering Problems by Dynamical Analogies*, Van Nostrand Reinhold, New York, 1966. Provides the means for deriving and solving the dynamical analogies considered in this chapter.

RIAA Standard Reproducing Characteristic, Record Industry Association of America, New York, 1964.

Tremaine, Howard M. *Audio Cyclopedia*, Howard W. Sams, Indianapolis, 1969.

White, James V. "A Linear Theory of Phonograph Playback," *Journal of the Audio Engineering Society*, Vol. 19, No. 2, p. 94, 1971.

(See also more than two hundred articles on disk phonograph recording and reproducing in the *Journal of the Audio Engineering Society*.)

10
Sound Motion
Picture Recorders
and
Reproducers

10.1 INTRODUCTION

A sound motion picture recorder consists of a light source modulated by an electro-mechanical modulator actuated by an audio frequency signal input. By that means it produces audio frequency modulated images upon the sound track of a moving motion picture film in which the sound track modulations correspond to the audio frequency signal input. A sound motion picture reproducer consists of a light source for illuminating the sound track of the moving film so that the light pulsations that pass through the sound track and fall upon the phototube produce an electrical signal output corresponding to the original audio frequency input in the recording. This chapter describes the processes and machines employed in sound motion picture recording and reproducing of audio frequency signals.

10.2 SOUND MOTION PICTURE RECORDING
PROCESS

A. Sound Motion Picture Recording System

The elements of a sound motion picture recording system are shown in Fig. 10.1. The first element is the acoustics of the set. The factors which influence

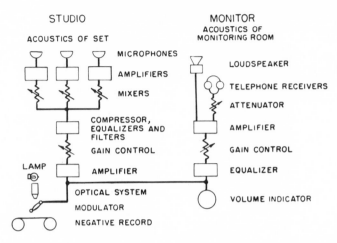

Fig. 10.1   The elements of a system for sound recording on motion picture film.

the collection of sound for sound motion pictures will be discussed in Section 15.4. The outputs of the microphones are amplified and fed to attenuators and termed mixers. If more than one microphone is used, as for example, when a soloist is accompanied by an orchestra, with one microphone for the soloist and one for the orchestra, the output of the two microphones may be adjusted for the proper balance. A low-pass filter is usually used to reduce ground noise above the upper limits of reproduction. A high-pass filter is used on speech with the lower limit placed below the speech range. This latter expediency reduces low frequency noises without impairing the speech quality. An equalizer is used to accentuate the high frequencies to compensate for the film transfer loss at high frequencies. A compressor is used to reduce the volume range (see Section 12.2). The following attentuator controls the over-all volume. The output of the amplifier feeds the light modulator and the monitoring system. By means of the optical system and light modulator, the electrical variations are recorded on the film into the corresponding variations in area, a process termed variable-area recording. (See Section 10.2B.) The monitoring system is also connected to the output of the recording amplifier. An equalizer is used to adjust the frequency characteristic to simulate that of the ultimate reproduction. If the monitoring is carried out in a room, a loudspeaker is used. When the monitoring and mixing is carried out on the set, earphones are used for monitoring.

During the past two decades in the recording of sound for sound motion pictures, the direct optical recording of the sound on film has been replaced by first recording on a magnetic tape master by the system shown in Fig. 10.2. The procedures in recording on magnetic tape for sound motion pictures are similar to those for recording on photographic film and will not be repeated.

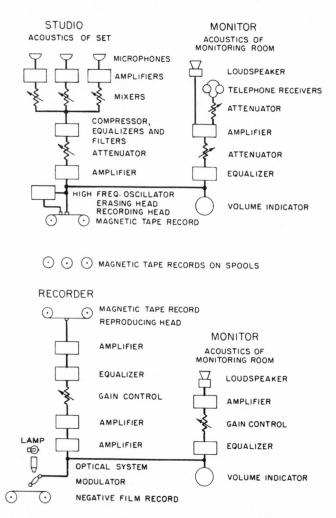

Fig. 10.2 The elements of the systems for a complete magnetic tape sound recording and reproducing system and a sound motion picture film recording system.

The system for recording the magnetic tape master will be described in Section 10.5. The magnetic tape sound record is reproduced and the output recorded on photographic film by the system shown in Fig. 10.2.

B. Recording Galvanometer

In the variable-area sound track the transmitted light amplitude is a function of the amount of unexposed area in the positive print. This type of sound track is produced by means of a mirror galvanometer that varies the width of the light slit under which the film passes. The elements of a variable-area recording system are shown in Fig. 10.3. The triangular aperture is uniformly illuminated by means of a lamp and lens system. The image of the triangular aperture is reflected by the galvanometer mirror focused on the mechanical slit. The mechanical slit in turn is focused on the film. The galvanometer mirror swings about an axis parallel to the plane of the paper. The triangular light image on the mechanical slit moves up and down on the mechanical slit. The result is that the width of the exposed portion of the negative sound track corresponds to the rotational vibrations of the galvanometer. In the positive record the width of the unexposed portion corresponds to the signal.

The amount of ground noise produced is proportional to the exposed

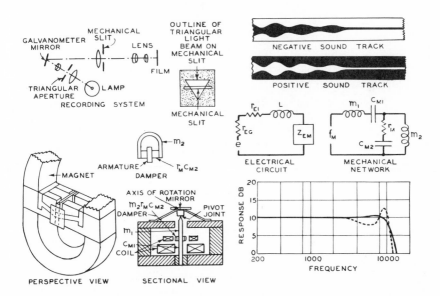

Fig. 10.3 The elements of a variable area sound motion picture film recording system. The negative and positive sound tracks. Perspective and sectional views, mechanical network, and electrical circuit of the galvanometer. In the mechanical network, f_M = the mechanical driving force. m_1 and C_{M1} = the mass and compliance of the armature. m_2, r_M, and C_{M2} = the mass, mechanical resistance, and compliance of the damper. In the electrical circuit, z_{EM} = the electrical motional impedance. L and r_{E1} = the damped inductance and electrical resistance. r_{EG} = the electrical resistance of the generator. e = the voltage of the generator. The graph depicts the frequency response characteristic of the galvanometer. Dotted and solid lines depict the frequency response for the galvanometer alone and with an electrical capacitance in shunt with the galvanometer, respectively.

portion of the positive sound track. For this reason it is desirable to make the unexposed portion of the record just wide enough to accommodate the modulation. This is accomplished by applying a bias signal to the galvano-meter. In the absence of a signal a very narrow exposed portion is produced on the negative record which means a correspondingly narrow unexposed portion on the positive record. When a signal appears, the triangular light spot on the mechanical slit moves down just enough to accommodate the signal. The initial bias is accomplished within a millisecond. However, the return to normal bias after a large signal followed by a small signal is about 1 second. Faster return action produces thumping in the reproduced record.

A film sound-reproducing system is an amplitude system; that is, the voltage output is proportional to the amplitude on the film. Therefore, in order to obtain a uniform frequency response characteristic, neglecting the frequency discrimination due to finite recording and reproducing slits, the amplitude of the galvanometer should be independent of the frequency. Perspective and sectional views, the electrical circuit, and the mechanical network of a film-recording galvanometer are shown in Fig. 10.3. The magnetic drive system is a balanced armature that moves the pivot point which in turn rotates the mirror. The controlling element in the vibrating system in the low frequency range is the compliance, C_{M1}. Under this condition the ratio of the amplitude to the applied force is independent of the frequency. A damper, m_2, r_M, C_{M2}, reduces the amplitude in the region of the resonant frequency of m_1 with C_{M1}. The frequency response characteristic is shown in Fig. 10.3. It will be seen that the rotational amplitude is uniform with respect to frequency to about 10,000 cycles.

C. Sound Track

The negative and positive sound tracks on motion picture film are depicted in Fig. 10.4. On 35-millimeter film the sound track occupies about 0.1 inch just inside the sprocket holes.

The sound track shown in Fig. 10.4 is termed the variable-area system, in which the transmitted light is a function of the unexposed area of the positive print. There is also the variable-density system, in which the light amplitude is an inverse function of the amount of exposure in the positive print. In recent years the variable-area system has replaced the variable-density system. The reproducer can play either one.

D. Recording Film-Transport Mechanism

The film-transport mechanism used in recording sound on film consists of a positive drive of the perforated film and a constant-speed drive of the film

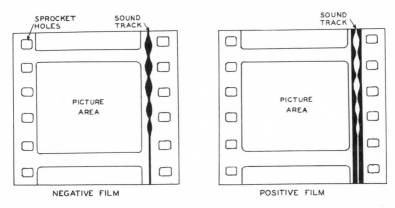

Fig. 10.4 The position of the picture and sound track on 35-millimeter sound motion picture film. A. Negative. B. Positive.

where the modulated light beam strikes the film. A film-transport mechanism of this type is shown in Fig. 10.5. Positive drive of the film is obtained by the sprocket drive. The sprocket drive is interlocked with the camera drive so that synchronism of the picture and sound will be obtained. (See Section 15.4.) When the film passes over the sprocket drive, variations in the motion of the film at the sprocket-hole frequency are produced. These variations in

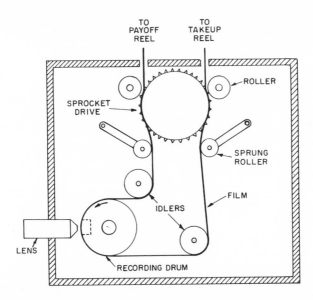

Fig. 10.5 Schematic view of the photographic film transport mechanism of a motion picture film sound recorder.

the film speed must be removed at the recording point to eliminate spurious frequency modulation of the image on the film. Uniform speed at the recording point is provided by the filter between the sprocket drive and the recording point, which consists of the inertia of the recording drum and the compliance of the film between the recording drum and the sprocket drive. The recording drum is driven by a magnetic system from the motor, which drives the sprocket and also supplies a slight amount of drive to the film. This form of drive furnishes isolation from variations in the rotational speed of the motor drive. The combination of the isolating filter and the magnetic drive provides a system with very uniform motion at the surface of the drum. The image of the modulator is focused on the film while it is in contact with the drum.

E. Recording of a Sound Motion Picture

The techniques employed in the recording of a sound motion picture on a sound stage are described in Section 15.4 and depicted in Fig. 15.9.

10.3 SOUND MOTION PICTURE FILM PRODUCTION PROCESS

The processes in the film laboratory for the mass production of sound motion picture positive prints are shown in Fig. 10.6. The negative record with picture and sound is developed as shown in Fig. 10.4. Then the required

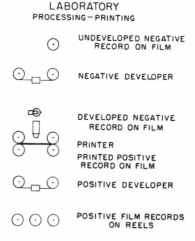

Fig. 10.6 The steps in the process for the production of motion-picture positive film from the negative film.

number of positive prints of both picture and sound are printed from the negative record as shown in Fig. 10.6. These positive prints are developed. These positive sound motion picture prints are used for the sound reproduction and picture reproduction in the theater.

10.4 SOUND MOTION PICTURE REPRODUCING PROCESS

A. Sound Motion Picture Reproducing System

The elements of a sound motion picture reproducing system are shown in Fig. 10.7. The variable-area sound track on the positive sound motion picture film depicted in Fig. 10.4 is reproduced in the theater by pulling the film past a slit illuminated by a light and a suitable optical system. The resultant variations in light, due to the variable area on the film, fall upon the phototube and are converted into the corresponding electrical variations. These are then amplified and fed to equalizers and filters. A low-pass filter is used to cut out the ground noise due to film above the upper limit of reproduction. An equalizer is used to adjust the frequency characteristic to that suitable for the best reproduction in the theater. The gain control is used for adjusting the level of reproduction. The output of the power amplifier feeds the stage loudspeakers and monitoring loudspeaker. The monitoring loudspeakers and the attenuator are located in the projection booth. As a matter of fact, the entire system, save for the stage loudspeakers, is located in the projection booth. A dividing network and a two-channel loudspeaker system are shown

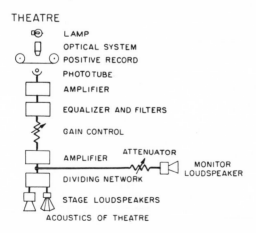

Fig. 10.7   The elements of a system for reproducing sound motion picture film.

in Fig. 10.7. Of course, any type of suitable loudspeaker described in Chapter 4 may be used. The action of a sound motion picture reproducer in a theater will be discussed in Section 16.6 and will not be repeated here.

B. Optical-Electronic Sound Motion Picture Reproducer

The elements of the optical-electronic sound motion picture reproducer are shown in Fig. 10.8. The light source, in the form of an incandescent lamp, is focused upon a mechanical slit by means of a condensing lens. The mechanical slit in turn is focused on the negative film. The height of the light-slit image on the film is usually about 0.00075 inch. Under these conditions the amount of light which impinges upon the phototube is proportional to the unexposed portion of the sound track in variable-area recording or to the inverse function of the density in variable-density recording. When the film is in motion, the light undulations which fall upon the phototube correspond to the voltage variations applied to the recording galvanometer. The voltage output of the phototube is proportional to the amount of light which falls upon the cathode. The voltage output frequency response characteristic of a typical motion picture film sound-reproducing system using a constant amplitude film is shown in Fig. 10.8. The falling off in response at the high frequency portion of the range is due to the finite dimensions of the slits in the recording and reproducing systems. This reduction in response can be overcome by frequency compensations in the recording and reproducing systems.

C. Reproducing Film-Transport Mechanism

The film transport used in reproducing sound on photographic film consists of a positive drive of the perforated film and a constant-speed drive where the light passes through the film to the phototube. A film-transport mechanism of this type is shown in Fig. 10.9. Positive drive of the film is obtained by means of the two sprocket drives. The sprocket drives are geared with the positive picture drive so that a constant loop of film is maintained between

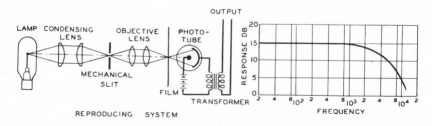

Fig. 10.8 The elements of a sound motion picture film reproducing system and the frequency voltage response characteristic with constant amplitude film.

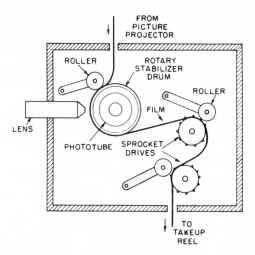

Fig. 10.9   Schematic view of the photographic film transport mechanism of a motion picture film sound reproducer.

the sound head and the picture head. The positive drive also insures that the film speed in reproduction will be same as that in recording. Since there is a loose loop of film between the picture head and sound head, variations in the picture drive will not be imparted to the sound head. After the film enters the sound head, it passes over a drum. The light beam of the reproducing system passes through the film to the photocell located inside the drum while the film is on the drum. Since the drum is rotated at a constant speed, the film will move past the light beam at a constant speed. The drum is driven by the first sprocket drive. The second sprocket isolates the takeup reel from the reproducing system. Uniform speed at the reproducing point is provided by the filter between the sprocket drive and the reproducing point, which consists of the inertia of the drum and the compliance of the film between the drum and the sprocket drive. To provide damping and stabilization, the drum drives a flywheel through a fluid coupling; the drum is termed a rotary stabilizer drum. The combination of the filter and the rotary stabilizer provides a system with very uniform motion at the surface of the drum.

10.5 SOUND MOTION PICTURE MAGNETIC TAPE RECORDING AND REPRODUCING PROCESSES

A. General Considerations

During the past two decades the original recording of sound directly upon motion picture film has to a large extent been replaced by recording on magnetic tape. The magnetic system for recording on magnetic tape is the same as that described in Section 8.2. However, in order to maintain sychronism between the picture recorded by the camera and the sound recorded on the magnetic tape, perforated tape as shown in Fig. 10.10A, sometimes referred to as magnetic film, is used in the sound recorder.

During the past two decades, wide-screen motion picture systems with stereophonic sound have been introduced on a wide scale. The information is rerecorded on the magnetic tracks cemented to the positive picture release print. The magnetic system for reproducing the magnetic tracks is the same as that described in Section 8.2. It is the purpose of this section to describe the magnetic tape, the tape transport used in recording, and the tape transport used in reproducing.

The camera and sound recorder must be interlocked as in the case of recording on photographic film. Therefore, a magnetic coating on a perforated plastic base is used as the recording medium. The dimensions are the same as those of 35-millimeter photographic film. One reason for the use of magnetic tape of these dimensions is that, if it is desired, the existing photographic recorder may be used by the addition of magnetic heads. Special magnetic recorders have also been developed, but these also use magnetic tape of the same dimensions. Another reason for the use of the wide tape is that several tracks representing several channels may be recorded on the tape shown in Fig. 10.10A because the total width of the magnetic coating is approximately an inch. For example, a soloist may be recorded on one track and the sections of the orchestra may be recorded on other tracks. The three or more channels of stereophonic sound may be recorded on the magnetic tape.

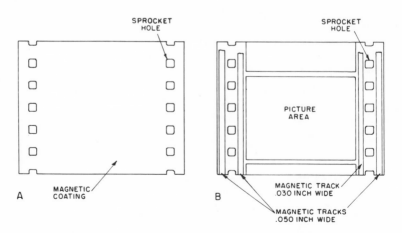

Fig. 10.10 A. Magnetic tape for original sound recording in sound motion pictures. B. Magnetic tracks on a motion picture positive film.

During the past two decades, wide-screen motion picture systems with stereophonic sound have been introduced on a large scale. The three or four channels are recorded on the tape shown in Fig. 10.10A. The information is rerecorded on the magnetic tracks cemented to the positive picture release print as shown in Fig. 10.10B.

B. Recording Magnetic-Tape-Transport Mechanism

The tape transport used in recording sound on magnetic tape for sound motion pictures consists of a positive sprocket drive of the perforated film and a constant-speed drive where the magnetic recording head is in contact with the tape. A magnetic-tape-transport-mechanism of this type is shown in Fig. 10.11. Positive drive of the tape is obtained by means of the sprocket drive. The sprocket drive is interlocked with the camera drive so that synchronism of the picture and sound will be obtained. (See Section 15.4.) When the tape passes over the sprocket drive, variations in the motion of the film at the sprocket-hole frequency are produced. These variations in the film speed must be removed at the recording point to eliminate spurious frequency modulation of the magnetic recording on the magnetic tape. Uniform speed at the recording point is provided by the mechanical filter between the sprocket drive and the recording point, which consists of the inertia elements of the two rollers and the two drums and the compliance of the tape between the sprocket drive and the rollers and the drums. The drums are equipped with flywheels to provide additional inertia. Damping of the inertia and compliance system is provided by the mechanical resistance of the dash pot. The spring system maintains a tight loop for the magnetic tape. This mechanical-system

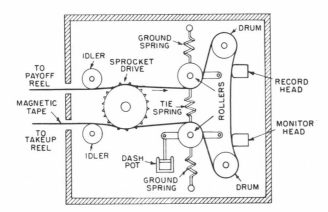

Fig. 10.11 Schematic view of the magnetic tape transport mechanism of a magnetic sound recorder for sound motion pictures.

design provides uniform motion at the magnetic heads. Two magnetic heads are provided, one for recording and the other for monitoring.

The film recorder of Fig. 10.5 may also be used for recording one track of magnetic tape by placing a magnetic recording head inside the drum in contact with the overhanging portion of the tape on the drum. The magnetic head is shown dotted in Fig. 10.5.

C. Reproducing Magnetic-Tape-Transport Mechanism

The information on the tape in the magnetic system of sound reproduction is rerecorded on the magnetic tracks of the positive film. (See Fig. 10.10B.) The transport mechanism used in reproducing sound recorded on magnetic tracks on positive film consists of a positive drive of the perforated film and a constant-speed drive at the point of contact of the magnetic heads as in Fig. 10.12. Positive drive of the film is obtained by means of two sprocket drives. These two sprocket drives are geared to the positive picture drive so that a constant loop of film is maintained between the sound head and the picture head. The positive drive also insures that the film speed in reproduction will be the same as in recording. The inertia drum coupled to a flywheel provides the inertia element of the filter system. The film loop tension rollers provide the compliance elements in the filter system. Damping of the loop tension rollers provides the mechanical resistance element. The filter system removes the variations introduced by the sprocket drive and provides uniform motion of the film at the magnetic heads.

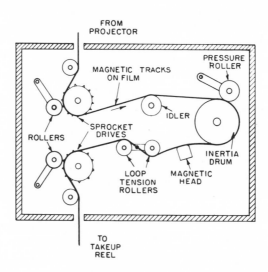

Fig. 10.12 Schematic view of the film transport mechanism of a magnetic reproducer for sound motion pictures.

D. Multichannel Sound Motion Picture System

The monophonic sound motion picture positive film employs the optical sound track depicted in Fig. 10.4. Processing is simplified because the positive picture and the positive sound track may be printed at the same time.

The stereophonic sound motion picture positive film employs three channels of stereophonic sound and another channel for the surround sound in the theater. The description of the system in the theater will appear in Section 16.6. The four sound channels on the positive film release print are carried on magnetic tracks as shown in Fig. 10.10B. The magnetic tracks are applied after the positive film has been printed. Then the sound is recorded on the four magnetic tracks. The four channels recorded on the four magnetic tracks are reproduced in the theater by means of the magnetic tape reproducer described in Section 10.5C together with the remainder of the equipment in each channel, as described in Section 10.4A.

10.6 SPEED AND SOUND TRACKS OF 35-, 16-, AND 8-MILLIMETER MOTION PICTURE FILMS

The recording of sound on 35-millimeter film has been described in this chapter. In the case of 35-millimeter film both optical and magnetic recording are used as described in this chapter. The speed of the 35 millimeter film is 18 inches per second.

The recording of sound on 16-millimeter film is carried out by either an optical or a magnetic recording system. The 16-millimeter film is a scaled-down version of the 35-millimeter film. In the positive-release prints there is one row of sprocket holes. The sound track is placed on the opposite side. The speed of the 16-millimeter film is eight inches per second. Therefore, the top frequency range that can be recorded is somewhat less than for the 35-millimeter film.

The recording of sound on 8-millimeter film is carried out by a magnetic recording system. There is only one row of sprocket holes. The magnetic track is placed opposite the sprocket holes. The speed of the 8-millimeter film is four inches per second.

REFERENCES

Olson, Harry F. *Acoustical Engineering*, Van Nostrand Reinhold, New York, 1957.
Tremaine, Howard M. *Audio Cyclopedia*, Howard W. Sams, Indianapolis, 1969.
(See also numerous articles on sound motion picture technology and standards in the *Journal of the Society of Motion Picture and Television Engineers*.)

11
Radio Broadcast
Transmitters
and
Receivers

11.1 INTRODUCTION

A radio transmitter consists of an electronic system whereby an audio frequency signal input modulates a radio frequency signal which is radiated into space. A radio receiver consists of an electronic system which receives and selects a small amount of an audio frequency modulated ratio frequency signal from the space radiated by a radio transmitter and converts the audio frequency modulated radio frequency signals to the original audio frequency signals. There are two types of radio broadcast systems, namely, amplitude modulation and frequency modulation. The amplitude modulation radio broadcast systems are monophonic, while the frequency modulation radio broadcast system is both monophonic and stereophonic. The monophonic frequency modulation system is used in television sound reproduction. The purpose of this chapter is to describe amplitude modulation and frequency modulation radio broadcast transmitters and receivers.

11.2 AMPLITUDE MODULATION RADIO BROADCAST SYSTEM

A. Monophonic Amplitude Modulation Radio Broadcast Transmitter

The elements of a monophonic amplitude modulation radio broadcast transmitter are shown in Fig. 11.1. The original source of sound signals may be the outputs of microphones in the studio and the control room, a disk phonograph reproducer, or a magnetic tape reproducer, or all three, as depicted in Fig. 11.1. The outputs of the microphones are amplified and applied to mixers. The outputs of the disk phonograph and magnetic tape

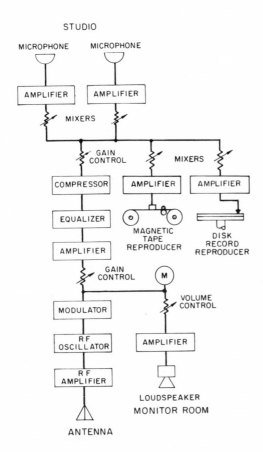

Fig. 11.1   The elements of a monophonic amplitude modulation broadcast radio transmitter.

reproducers are also applied to the mixers. The microphones are located in the studio and the control room as described in Section 15.2 and depicted in Fig. 15.1. A soundproof glass wall partition which separates the studio and the control room gives the engineer a full view of the action in the studio. (See Fig. 15.1.) The mixers, gain controls, and volume indicators constitute the console located in the monitoring room. The disk phonograph reproducer and magnetic tape reproducer are located in the monitoring room. (See Fig. 15.1.) The output of the mixers is fed to a gain control that is followed by a compressor. (See Section 12.2.) The compressor is followed by an equalizer which compensates for any frequency variation in the overall system. There is another gain control, which governs the input to the modulator. The monitoring system, consisting of a volume control, an amplifier, and a loudspeaker and volume indicator is connected at the output of the gain control. The monitoring system is located in the monitoring room. (See Section 15.2 and Fig. 15.1.) The modulator provides the means for modulating the amplitude of the oscillator to correspond to the audio frequency signal input to the modulator. The output of the oscillator is amplified by the radio frequency amplifier which in turn excites the antenna. The audio frequency, amplitude-modulated radio waves are radiated into space by the antenna.

B. Monophonic Amplitude Modulation Radio Broadcast Receiver

The elements of a monophonic amplitude modulation broadcast radio receiver are shown in Fig. 11.2. A very small portion of the radio frequency radiated by the transmitter antenna is picked up by the receiving antenna. The output of the receiving antenna is amplified by a tuned radio frequency amplifier. The output is combined with that of a higher radio frequency oscillator and fed to the first detector. The resultant intermediate-frequency carrier, with the original modulation, is amplified by the intermediate-frequency amplifier. The second detector converts the modulated intermediate-frequency carrier to an audio frequency electrical wave in which the variations correspond to the undulations in the original sound wave at the transmitter. The second detector is followed by a volume control, which controls the level of the reproduced sound in the room. The volume control is followed by a power amplifier which drives the loudspeaker. The loudspeaker converts the electrical variations into the corresponding sound vibrations. The latter vibrations correspond to the original sound waves in the studio. The radio receiver is usually operated in a small room, as for example, the living room in the home. (See Section 16.3 and the systems depicted in Figs. 16.2, 16.3, 16.4, and 16.5.)

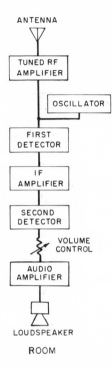

ANTENNA

TUNED RF
AMPLIFIER

OSCILLATOR

FIRST
DETECTOR

IF
AMPLIFIER

SECOND
DETECTOR

VOLUME
CONTROL

AUDIO
AMPLIFIER

LOUDSPEAKER

ROOM

Fig. 11.2   The elements of a monophonic amplitude modulation broadcast radio receiver.

C. Performance Characteristics of a Monophonic Amplitude Modulation Radio Broadcast System

1. Transmitter. The main performance characteristics of a monophonic amplitude modulation radio broadcast transmitter are as follows:

The radio frequency spectrum covered is 535 to 1605 kilohertz.

The frequency range of the audio applied to the amplitude modulator is 30 to 15,000 hertz within ± 1 dB.

The signal-to-noise ratio is more than 60 dB.

The nonlinear distortion is less than 1 percent harmonic nonlinear distortion at 90 percent modulation.

2. Receiver. The main performance characteristics of a high-quality amplitude modulation radio broadcast receiver are as follows:

The radio frequency spectrum covered is 535 to 1605 kilohertz.

The frequency response characteristic of such a receiver is that shown in Fig. 11.3. The separation between transmitters is 10 kilohertz. Therefore,

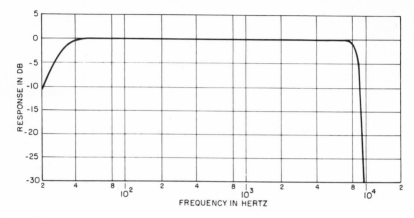

Fig. 11.3 The audio frequency response characteristic of a high quality monophonic amplitude modulation broadcast radio receiver.

there will be a 10-kilohertz tone if the two transmitters provide approximately the same field strength at the receiver. In order to eliminate this tone there should be high attentuation at 10 kilohertz. The frequency response characteristic shown in Fig. 11.3 is for a high-quality receiver. In commercial receivers the frequency response covers the frequency range of Fig. 11.3 all the way down to a range of 500 to 2000 hertz for the pocket-type radio receiver.

The sensitivity is 15 microvolts for a 30-dB signal-to-noise ratio.

The signal-to-noise ratio is 60 dB for high signal inputs.'

The nonlinear harmonic distortion is less than 2 percent.

11.3 FREQUENCY MODULATION RADIO BROADCAST SYSTEM

A. Monophonic Frequency Modulation Radio Broadcast Transmitter

The elements of a monophonic frequency modulation broadcast transmitter are shown in Fig. 11.4. The original source of sound signals may be the outputs of the microphones in the studio and the control room, a disk phonograph reproducer, or a magnetic tape reproducer, or all three, as depicted in Fig. 11.4. The outputs of the microphones are amplified and applied to mixers. The outputs of the disk phonograph and magnetic tape reproducers are also applied to mixers. The microphones are located in the

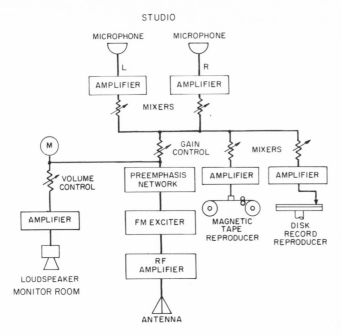

Fig. 11.4 The elements of a monophonic frequency modulation broadcast radio transmitter.

studio and control room as described in Section 15.2 and depicted in Fig. 15.1. A soundproof glass wall partition which separates the studio and control room gives the engineer the full view of the action in the studio. The mixers, gain controls, and volume indicators constitute the console located in the monitoring room. The disk phonograph reproducer and magnetic tape reproducer are located in the monitoring room. (See Fig. 15.1.) The output of the mixers is fed to a gain control. The monitoring system, consisting of a volume control, amplifier, and loudspeaker, and a volume indicator, is connected at the output of the gain control. The monitoring system is located in the monitoring room. The output of the gain control is applied to the preemphasis network. The preemphasis network exhibits a frequency response characteristic given by:

$$R = 20 \log_{10} \sqrt{1 + 0.22 \times 10^{-6} f^2} \qquad (11.1)$$

where $R =$ response, in decibels, and
$f =$ audio frequency, in hertz.

The purpose of the preemphasis is to increase the signal-to-noise ratio in transmission. The preemphasis network is followed by the frequency modulation exciter. The frequency modulation exciter consists of either a reactance tube or the more modern varactor diode and network which frequency modulates a radio frequency oscillator in correspondence with the audio frequency input. The output of the exciter drives the radio frequency amplifier. The output of the radio frequency amplifier excites the antenna. The audio frequency, frequency modulated radio waves are radiated into space by the antenna.

B. Monophonic Frequency Modulation Radio Broadcast Receiver

The elements of a monophonic frequency modulation radio broadcast receiver are shown in Fig. 11.5. A very small portion of the radio frequency radiated by the transmitter antenna is picked up by the receiving antenna. The output of the receiving antenna is amplified by a tuned radio frequency amplifier termed the tuner of Fig. 11.5. The output of the tuner is combined

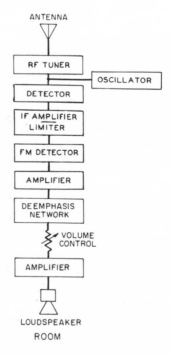

Fig. 11.5 The elements of a monophonic frequency modulation broadcast radio receiver.

with the output of a radio frequency oscillator and fed to a detector. The resultant intermediate frequency carrier with the original frequency modulation is amplified and limited. The limiter is an integral part of the intermediate frequency amplifier. The limiting provides a reduction in noise. The output of the limiter is fed to a frequency modulation detector. The frequency modulation detector consists of a frequency discriminator that converts the frequency modulated intermediate frequency carrier to an audio frequency wave in which the variations correspond to the undulations in the original sound wave at the transmitter. The detector is followed by the deemphasis network, which is the inverse of the preemphasis network at the transmitter. That is, the deemphasis network provides a frequency response characteristic which is the inverse of that given by Equation (11.1). The high frequency discrimination of the preemphasis network provides an increase in the signal-to-noise ratio. The deemphasis network is followed by a volume control, which controls the intensity of the reproduced sound in the room. The volume control is followed by the power amplifier, which drives the loudspeaker. The loudspeaker converts the audio-electrical variations into the corresponding sound vibrations. The latter vibrations correspond to the original sound waves in the studio. The radio receiver is usually operated in a small room, as for example, the living room in the home. (See Section 16.3 and Figs. 16.2, 16.3, 16.4, and 16.5.)

C. Television Audio Transmitter and Receiver

The television audio system is of the monophonic frequency modulation type. The audio and video sections are combined in the television transmitter. The audio receiver consists of the audio and video sections. The operation of the audio section of the transmitter and receiver is similar to that of the monophonic frequency modulation system.

D. Stereophonic Frequency Modulation Radio Broadcast Transmitter

Stereophonic frequency modulation radio broadcast provides for the transmission of the sum of the left and right stereophonic audio signals as the direct-frequency modulation of the radiated carrier as depicted by the spectrum diagram of Fig. 11.6. The difference between the left and right stereophonic audio signals is used to amplitude-modulate a 38-kilohertz subcarrier in suppressed-carrier fashion, with the result impressed as frequency modulation on the radiated carrier as shown in Fig. 11.6. In addition, a pilot subcarrier of 19 kilohertz, shown in Fig. 11.6, is transmitted for synchronizing the FM receiver.

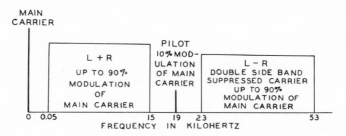

Fig. 11.6 Spectrum of the signals of a stereophonic frequency modulation broadcast radio system.

The elements of a stereophonic frequency modulation broadcast transmitter are shown in Fig. 11.7. The stereophonic inputs consist of microphones in the studio and the control room, a stereophonic disk phonograph reproducer, and a stereophonic magnetic tape reproducer. The monitoring equipment is the same as for the monophonic system described in the preceding sections for the monophonic amplitude and frequency modulation receivers except that there are two channels. The output of the mixing system is fed to a matrix. The output of the matrix contains two audio signals, the $L + R$ and the $L - R$. Both are passed through preemphasis networks as described above in Section 11.3A. The $L + R$ preemphasized signal is applied to the 58-kilohertz low-pass filter. The $L - R$ preemphasized audio signal is applied to a 38-kilohertz AM double-sideband modulator in which the 38-kilohertz carrier is suppressed. A 19-kilohertz signal is produced by a pilot generator. The suppressed-carrier modulation signal of $L - R$ and the pilot signal are combined and applied to the 58-kilohertz low-pass filter. The output of the low-pass filter is fed to the frequency modulation exciter that was described in Section 11.3A. The output of the exciter drives the radio frequency amplifier. The output of the radio frequency amplifier excites the antenna. The stereophonic audio frequency, frequency modulated radio waves are radiated into space by the antenna.

E. Stereophonic Frequency Modulation Radio Broadcast Receiver

The elements of a stereophonic frequency modulation broadcast receiver are shown in Fig. 11.8. The tuner, radio frequency oscillator, detector, and intermediate frequency amplifier are similar to the monophonic receiver described in Section 11.3B. The output of intermediate-frequency amplifier is detected by a frequency modulation detector, the output of which is

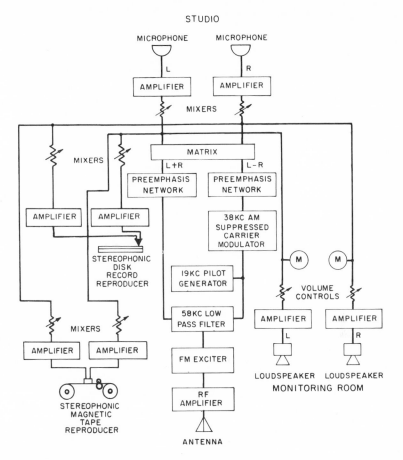

Fig. 11.7 The elements of a stereophonic frequency modulation broadcast radio transmitter.

applied to a 15-kilohertz low-pass filter, a 23-to-53-kilohertz band-pass filter, and a 19-kilohertz narrow-band filter. The output of the 19-kilohertz filter is coupled to a 38-kilohertz doubler oscillator which generates the 38-kilohertz carrier. The carrier is combined with the two sideband signals from the 23-to-53-kilohertz filter and fed to an amplitude modulation detector. The resultant output of the amplitude modulation detector is the $L - R$ audio signal. The output of the 15-kilohertz low-pass filter contains the $L + R$ audio signal. The $L - R$ and $L + R$ signals are fed to a matrix, the output of which provides the L and R audio signals. The deemphasis network

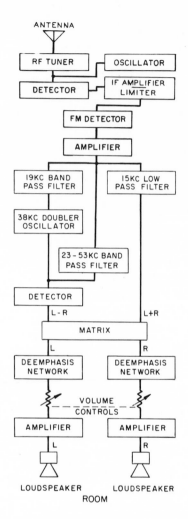

Fig. 11.8 The elements of a stereophonic frequency modulation broadcast radio receiver.

normalizes the response with respect to frequency of the audio signals. The audio signals are amplified and converted to the left and right sound signals in air by the loudspeakers. The stereophonic frequency modulation radio receiver shown in Fig. 11.8 is usually operated in a room in the home as described in Section 16.3 and depicted by the systems of Figs. 16.2, 16.3, 16.4, and 16.5.

F. Performance Characteristics of Monophonic and Stereophonic Frequency Modulation Radio Broadcast Systems

1. Transmitter. The main performance characteristics of monophonic and stereophonic radio broadcast transmitters are as follows:

The radio frequency spectrum covered is 88 to 108 megahertz.

The frequency range of the audio applied to the frequency modulator is 30 to 15,000 hertz within ± 1 dB.

The signal-to-noise radio is more than 60 dB.

The nonlinear distortion is less than 1 percent harmonic nonlinear distortion for 100 percent modulation.

2. Receiver. The main performance characteristics of a high-quality monophonic or stereophonic frequency modulation radio broadcast receiver are as follows:

The radio frequency spectrum covered is 88 to 108 megahertz.

The frequency response characteristic for both monophonic and stereophonic receivers is that shown in Fig. 11.9A. The channel cross-talk separation frequency response between channels for a stereophonic receiver is shown in Fig. 11.9B. The frequency response shown in Fig. 11.9 is for a high-

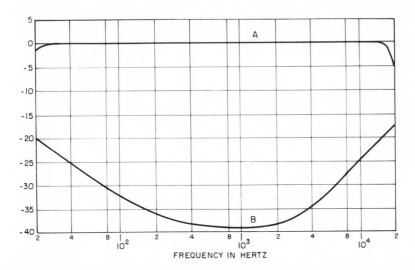

Fig. 11.9   A. The audio frequency response characteristic of a high quality frequency modulation broadcast radio receiver. B. Audio channel cross-talk separation frequency response for a frequency modulation broadcast radio receiver.

quality receiver. In commercial receivers the frequency response covers the frequency range shown in Fig. 11.9 all the way down to a range of 500 to 2000 hertz for a pocket-type radio receiver.

The sensitivity for a high-quality receiver is 2 microvolts.

The signal-to-noise ratio is more than 60 dB for high signal inputs.

Nonlinear harmonic distortion is 0.5 percent for monophonic receivers and less than 1 percent for stereophonic receivers.

REFERENCE

McGraw-Hill Encyclopedia of Science and Technology, McGraw-Hill, New York, 1971.

12
Accessories
to Sound
Reproduction

12.1 INTRODUCTION

The preceding chapters have provided expositions on the fundamental audio transducers employed in sound-reproducing systems, as for example, microphones, amplifiers, earphones, and loudspeakers. There are other audio transducers that serve as accessories to sound reproduction, namely, compressors, limiters, noise reducers, frequency response correctors, reverberators, and delayers. These accessories are used to improve and enhance the performance of sound reproduction. The purpose of this chapter is to provide descriptions and expositions of the preceding list of accessories to sound reproduction.

12.2 COMPRESSOR AND LIMITER

A volume compressor is an electronic system that reduces the amplification when the signal input attains a certain level. Compressors are used to reduce the volume or amplitude range in sound motion picture, magnetic tape and phonograph disk recording, sound broadcasting, sound-reenforcing systems, and so on.

A volume compressor consists of a variable-gain amplifier in which the gain is a function of the general level of the signal. The elements of a compressor are shown in Fig. 12.1. The variable-gain stage may be a field effect transistor in which the control voltage e is applied to the gate. The output of the variable-gain amplifier is amplified and rectified. The direct current is used to charge a capacitor. Up to a certain point there is a constant relationship between the input and output voltages, as is depicted in the graph of Fig. 12.1. Beyond this point the control voltage e attains a value which reduces the gain of the variable-gain amplifier. (A field effect transistor may be used as a variable-gain amplifier.) As the voltage e increases beyond this point, the voltage gain continues to decrease. The attack time, that is, the charging of the capacitor C_E, is very rapid, being less than a millisecond. The decay time of the capacitor C_E shunted by the resistor r_{E1} is relatively long. The retreat to normal is of the order of a second.

Fig. 12.1 shows a gradual reduction in gain with increase of the input. A reduction in the volume range in radio, magnetic tape, and phonograph sound reproduction makes it possible to reproduce the wide amplitude range of orchestra music in the home without excessive top levels. The use of the compressor improves the signal-to-noise ratio and thereby improves the intelligibility of speech and enhances the music sound reproduction when the ambient noise is high.

A limiter is similar to the compressor except that the relationship between the input and output is constant up to a certain input. Beyond that point the output does not increase. The limiter is useful for protection against sudden overloads in the sound-reproducing system.

12.3 NOISE REDUCTION BY COMPRESSION AND EXPANSION

For decades compression has been in use in recording, with or without complementary expansion in sound reproduction. However, if there is

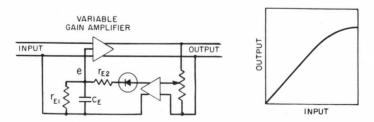

Fig. 12.1 The elements of an audio compression system. The graph depicts the input-output characteristic of the compressor.

complete restoration of the amplitude characteristic in reproducing by complementary expansion, there are some extraneous sounds introduced in the process. The most troublesome is the "breathing" or "pumping" action of the compressor and limiter, which is due to a raising and lowering of the background noise as the gain of the variable-gain amplifier changes.

The "breathing" and "pumping" problems have been solved by dividing the audio frequency range into frequency bands and applying separate compression and complementary expansion to each band as depicted in Fig. 12.2. The audio frequency range is divided into four frequency bands. There is a limiter which limits the input to the compressor or the expander. There may be some momentary clipping of the signal, but since this occurs at high levels, the distortion is not apparent. In the compression system of Fig. 12.1 a variable-gain amplifier is used. In the compression system of Fig. 12.2 a differential network operates on the incoming signal and develops a similar signal that is combined with the input signal by means of the adder in such a manner that at certain levels a compression of the signal takes place. The compressor characteristic is shown in Fig. 12.2. In the expansion system

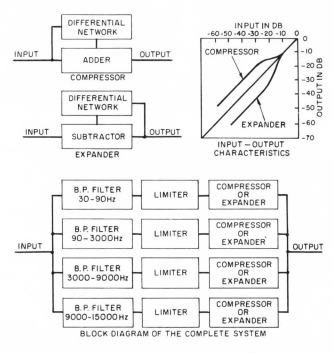

Fig. 12.2 Block diagrams of the compressor, expander and the complete system for recording by means of compression and expansion in reproducing and thereby reduce the noise. The graph depicts the input-output characteristics of the compressor and expander.

of Fig. 12.2, the same differential network operates on the signal except in this case the signal developed by the differential network is subtracted from the incoming signal as shown in Fig. 12.2. The expander characteristic is shown in Fig. 12.2. The graph of Fig. 12.2 shows that the compressor and the expander characteristics are complementary. The use of the compression system in recording on magnetic tape and expansion systems in reproducing reduces the noise in magnetic tape recording. A noise reduction of 10 dB can be obtained without any apparent distortion of any type.

12.4 THRESHOLD NOISE REDUCTION

The threshold noise reduction system consists of a nonlinear element which allows the useful signal to pass and discriminates against the noise. The threshold noise reduction system lowers the audio ground noise and thereby increases the signal-to-noise ratio.

A block diagram of the threshold noise reduction system is shown in Fig. 12.3. Band-pass filters which pass audio frequencies over the range of an octave are used at the input of the nonlinear element.

The nonlinear element consists of two solid state diodes connected as shown in Fig. 12.4. The input-output characteristic of the diode assembly is

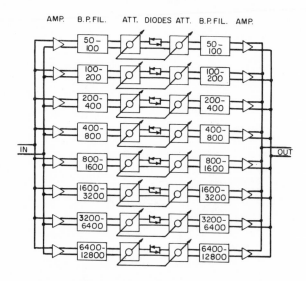

Fig. 12.3 Block diagram of a threshold noise reduction system employing a nonlinear element for discriminating against noise. The frequency ranges of the octave band-pass filters are in hertz.

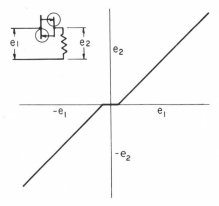

Fig. 12.4 The elements of a nonlinear system for discriminating against noise. The graph shows the input-output characteristic of the nonlinear element.

shown in Fig. 12.4. The input-output characteristic shows that for signals below a certain level there will be no output. This level in each octave is made to coincide with noise level by use of the complementary input-output level controls.

The output of the nonlinear element produces harmonics and subharmonics. However, since the pass band of the input and output band-pass filters is an octave, the harmonics and subharmonics will not be transmitted by the system. For example, if a sine wave is impressed upon the system, the output wave will be a sine wave because all harmonics and subharmonics of the sine wave will be rejected by the input and output band-pass filters. If two sine waves that are of different frequencies but confined to the octave by the input filter are impressed upon the nonlinear element, the harmonics, subharmonics, and sum and difference frequencies will be rejected by the output filter. A noise reduction of more than 20 dB can be obtained with the threshold system.

12.5 FREQUENCY RESPONSE MODIFIER

There are occasions where a modification of the overall frequency response characteristic may be desired. The most common example is the performance of a sound-reproducing system in a room in which the acoustics are not the desired optimum and as a result the frequency response is not the optimum for good reproduction of sound. For this application a frequency response modifier may be used to provide the optimum frequency response of the system.

In one form, the frequency response modifier consists of a bank of octave band-pass filters and amplifiers with means for adjusting the gain in each pass band as shown in Fig. 12.5. The frequency response characteristics of the band-pass filters are shown in Fig. 12.5A. When the gain is the same in all the channels, the overall response will be uniform with respect to frequency. The gain in each channel can be adjusted by means of the attenuator to provide all manner of frequency response characteristics, of which one is depicted in Fig. 12.5B. The dip in the response at 1000 hertz is 7 dB. However, a reduction in any octave up to 20 dB can be obtained.

A smaller number of broader or a larger number of narrower band-pass filters may be used, depending upon the desired application.

12.6 REVERBERATORS

The reverberation of studios may be changed and controlled within certain limits as described in Section 15.8. The amount of control that may be obtained by varying the amount of absorption by means of hard panels which cover the absorbing material is limited. Furthermore, the reproducing conditions may also require additional reverberation. Where the reverberation time of reproduced sound is far below the optimum value, sound reproduction may be enhanced by artificially adding reverberation.

Artificial reverberation may be added to the original sound signal by

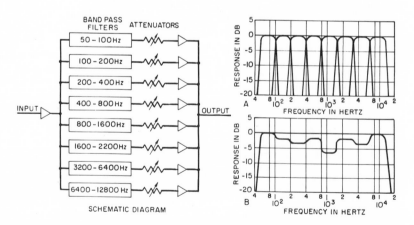

Fig. 12.5 Block diagram of a frequency response modifier. A. The frequency response characteristic of the octave elements. B. A typical frequency response characteristic which may be obtained by means of the frequency response modifier.

means of the amplifier, loudspeaker, reverberant chamber, microphone, and amplifier combination shown in Fig. 12.6A. The reverberant chamber consists of an enclosure with highly reflecting walls, ceiling, and floor. The cubical content varies from 1000 to 10,000 cubic feet. A maximum reverberation time of 5 seconds is possible. If a reduction in reverberation time is desired, flats of absorbing material may be brought into the chamber.

A two-dimensional reverberator consists of a thin plate of brass with amplified sending and receiving transducers as shown in Fig. 12.6B. The sending and receiving transducers are arranged for the production and reception of transverse vibrations of the plate. The transverse vibrations are propagated at a very low speed. Therefore, a relatively small plate simulates the dimensions of a room, except that instead of three-dimensional vibrations for the room the vibrations of the plate are two-dimensional. The reverberation time can be varied by means of the plate-and-screw damping arrangement. The close proximity of the damping plate to the brass plate damps the vibrations in the brass plate. The damping increases as the distance between the plates decreases. The reverberation time can be varied from 5 seconds to below 1 second by means of the damping system depicted in Fig. 12.6B.

A one-dimensional reverberator consists of a spring with amplified sending and receiving transducers attached at the two ends of the spring as shown in Fig. 12.6C. The velocity of propagation of vibrations in a spring is relatively low, being about 50 to 100 feet per second. Therefore, a relatively short spring will simulate the vibrations of a room in one dimension. The reverberation time can be varied from 5 seconds to below 1 second by means of the damping screw as depicted in Fig. 12.6C.

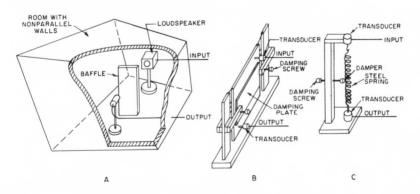

Fig. 12.6 Perspective views of reverberators. A. Loudspeaker-room-microphone reverberator. B. Plate type reverberator with input and output transducers. C. Spring type reverberator with input and output transducers at the two ends.

12.7 AUDIO DELAYERS

An audio delayer is an electro-acoustic system for introducing delays in sound reproduction system of up to 0.5 seconds or even more. The loudspeaker, pipe, and microphone combination and the magnetic tape reproducer are the most common systems for introducing delay in a sound reproduction system.

An audio delay system consisting of a power amplifier, loudspeaker, pipe, and microphone, and amplifier is shown in Fig. 12.7A. The audio delay is given by:

$$D = \frac{L}{1.1} \tag{12.1}$$

where D = delay, in milliseconds, and
       L = distance between the loudspeaker and microphone, in feet.

There is attenuation in the pipe, which increases with frequency. Amplitude compensation with respect to frequency is introduced in driving the amplifier to compensate for the frequency discrimination due to attentuation. The pipe beyond the microphone is damped with tufts of hair felt to absorb the sound waves beyond this point and thereby eliminate reflections at the end of the pipe. Delay of any reasonable value can be obtained by the assembly of units shown in Fig. 11.7A. (The use of the delay system will be described in Section 16.5.)

An audio delay system consisting of an audio magnetic tape reproducer with multiple magnetic heads is shown in Fig. 12.7B. The delay is given by

$$D = \frac{1000L}{V} \tag{12.2}$$

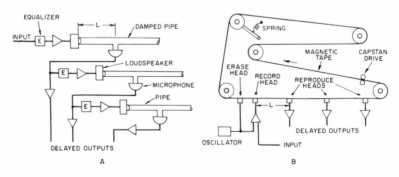

Fig. 12.7 Schematic views of audio delayers. A. Loudspeaker-pipe-microphone delayer. B. Magnetic tape delayer.

where D = delay, in milliseconds,
V = speed of the tape, in inches per second, and
L = distance between the record and reproduce head, in inches.

The magnetic tape delay system provides a simple means for introducing delay in an audio system. However, the objection is that the tape wears out after a few hours and must be replaced. Other parts also wear out, as for example, the heads. Delay of any reasonable value can be obtained by the magnetic tape system of Fig. 12.7B.

An audio delay system employing a digital delay system is depicted in the schematic diagram of Fig. 12.8. The first element is a low pass filter which eliminates all audio components above 12,000 hertz. The sampling switch operates at a frequency of 30,000 hertz. A section of the audio wave signal input is shown in Fig. 12.8A. The sampled audio signal is shown in Fig. 12.8B. The sampled signal is converted into the proper digital format by the analog to digital converter and fed into the digital shift register serving as a memory. The shift register elements are connected so that each element transfers its digital word to the next stage and reads the word from the previous stage during the sampling interval. In this way each word is entered at the input of the memory and moves down the line relatively slowly. The output of shift register is fed to the digital to analog converter and then to the sampling switch. The output of the sampling switch is shown in Fig. 12.8C with a delay indicated as t_D. The output of the sampling switch is smoothed by the low pass filter. The audio wave signal output with a delay of t_D seconds is shown in Fig. 12.8D which is the same as the audio wave input but with the delay of t_D seconds. The master synchronous controller synchronizes and controls all the functions with respect to time.

The signal to noise ratio is given by

$$\frac{S}{N} = 20 \log_{10} (2^n) \tag{12.3}$$

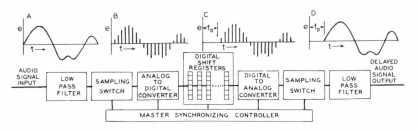

Fig. 12.8 The elements of an audio delay system employing a digital delay system. A. Audio wave signal input. B. Sampled audio digital signal. C. Delayed and sampled output signal. D. Delayed output audio signal.

where S/N = signal to noise ratio, in decibels, and
           n = number of bits in a digital word.

The storage requirements of the memory are given by

$$W = \frac{1}{3}\left(\frac{S}{N}\right)f_M t_D \tag{12.4}$$

where W = storage capacity, in bits,
        S/N = signal to noise ratio, in decibels,
        f_M = maximum audio frequency passed by the system, in hertz, and
        t_D = delay time, in seconds.

Delays in elements of 40 milliseconds are usually employed in sound reenforcing systems. However, delays of 0 to 320 milliseconds can be obtained. The signal to noise is usually about 60 dB. The audio bandwidth is 12,000 hertz.

The use of the delay system will be described in Section 16.5.

REFERENCES

Blesser, Barry. "Applying Digital Technology to Audio: Delay, Transmission, Storage, and Other Forms of Processing," Audio Engineering Society, Preprint No. 826, Convention, October 5–8, 1971.

Boner, C. R. "Some Examples of Sound-System Correction of Acoustically Difficult Rooms," Vol. 15, No. 2, p. 218, 1967.

Dolby, Ray M. "An Audio Noise Reduction System," *Journal of the Audio Engineering Society*, Vol. 15, No. 4, p. 383, 1967.

Olson, Harry F. *Acoustical Engineering*, Van Nostrand Reinhold, New York, 1957.

Olson, Harry F. "Audio Noise Reduction Circuits," *Electronics*, December 1947, p. 118.

Rettinger, Michael. "Note on Reverberation Chambers," *Journal of the Audio Engineering Society*, Vol. 5, No. 1, p. 108, 1957.

Tremaine, Howard M. *Audio Cyclopedia*, Howard W. Sams, Indianapolis, 1969.

(See also several articles on artificial reverberation systems in the *Journal of the Audio Engineering Society*.)

13
Acoustical
Measurements

13.1 INTRODUCTION

In sound reproduction, as in any field of applied science, theoretical analysis and analytical developments are substantiated by experimental verifications. In view of the importance of acoustical measurements in sound reproduction, it seems logical to devote a chapter to the subject of acoustical measurements as related to sound reproduction. To provide an exposition on all measurement techniques that have been developed in the field of sound reproduction would require an entire book. Therefore, this chapter must be confined to descriptions of the major and fundamental systems used in the measurement of the performance of the elements and systems employed in the reproduction of sound.

13.2 SOUND LEVEL METER

The basic instrument used in measurements in acoustics and sound reproduction is the sound level meter. The elements of a sound level meter are depicted in Fig. 13.1. The microphone calibrated in terms of a free sound wave should be essentially nondirectional. The attenuator and meter are

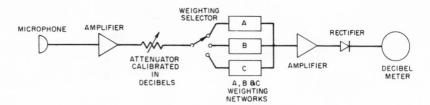

Fig. 13.1 Schematic diagram of the elements of a sound level meter.

calibrated in decibels above the threshold level of 0.0002 microbars at 1000 hertz. The overall response of an ideal sound level meter should be the reciprocal of the frequency response characteristic of the human hearing mechanism shown in Fig. 17.1. This would make the sound level meter unduly complex. Therefore, three frequency response characteristics have been standardized for the sound level meter as shown in Fig. 13.2. The *A* scale of Fig. 13.2 is the one selected for most sound level measurements. Sound levels measurements using this scale are designated as dBA.

There are some sound level measurement problems, particularly connected with the measurement of noise, where analysis with respect to frequency will provide additional information. One of the simplest frequency analyzers is the one in which the frequency band is divided into octaves. A schematic block diagram of a sound level meter incorporating an octave-band analyzer is shown in Fig. 13.3. The frequency response characteristics of the band-pass filters are shown in Fig. 13.4. The sound level is measured in each octave frequency band. The measurement may be made on the *A*, *B*, or *C* scales.

The sound level meter of Fig. 13.1 may be used in noise analysis in offices, factories, schools, restaurants, auditoriums, etc. The sound level meter may

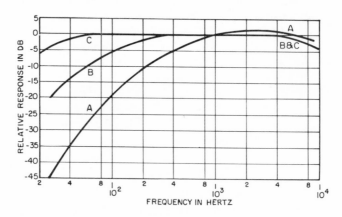

Fig. 13.2 The *A*, *B* and *C* frequency response characteristics of a sound level meter.

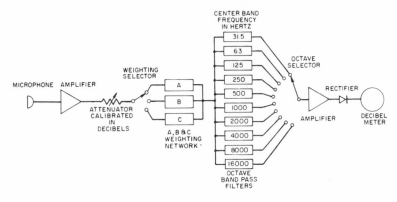

Fig. 13.3 Schematic diagram of the elements of a sound level meter with an octave frequency band analyzer.

also be used in determining the transmission or attenuation of sound by walls, ceilings, floors, windows, and doors. For more complex measurements of noise as, for example, that of machinery, the octave-band sound level meter of Fig. 13.3 will provide more information on the nature of the noise.

13.3 AUDIOMETER

The acuity of hearing is measured by means of an audiometer. The audiometer shown in Fig. 13.5 consists of an oscillator for generating pure tones,

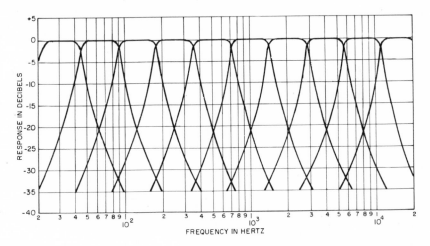

Fig. 13.4 The frequency response characteristic of the octave frequency band pass filters.

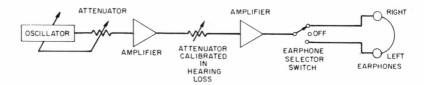

Fig. 13.5 Schematic diagram of the elements of an audiometer.

an attenuator calibrated in decibels relative to the relative threshold sound pressure, an amplifier, and an earphone. The test tones are 125, 250, 500, 1000, 1500, 2000, 3000, 4000, 6000, and 8000 hertz. In most measurements the values 1500, 3000, and 6000 hertz are not used. There are three normal reference threshold curves, designated as follows: ASA, Z24.5-1951, American Standards Association; ISO, 389-1964, International Standards Organization; ANSI, S3-6-1969, American National Standards Institute. The three thresholds are depicted in Fig. 13.6. The ANSI specification on threshold is the one now employed in audiometers. The audiometers now employ the Telephonics Type TDH-39 Earphone. The attenuator on the audiometer is calibrated in decibels above the threshold characteristic of Fig. 13.6. For a reading of 0 dB, the threshold of Fig. 13.6, there is no hearing loss. Any hearing threshold level above this constitutes a hearing loss.

The threshold of hearing is determined by a sine wave signal applied to one earphone or ear at a time. The sine wave signal is raised by means of the attenuator until it can just be heard. The reading in decibels on the attenuator shows the hearing loss.

In representing the hearing loss of a person, a convenient system is the graphed audiogram of Fig. 13.7. The audiogram shown by the curves A of

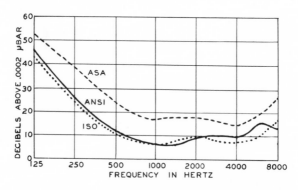

Fig. 13.6 The ASA, ISO, and ANSI standard reference threshold sound pressures employing a Telephonics Type TDH-39 Earphone. 0 dB = .0002 microbar.

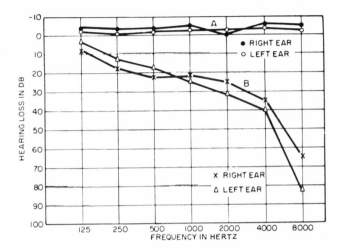

Fig. 13.7 Typical audiograms. A. Audiogram of a person with normal hearing. B. Audiogram of a person at the boundary of impaired hearing.

Fig. 13.7 are those for a person with normal hearing. The audiogram of a person at the boundary of impaired hearing is shown by the curves *B* of Fig. 13.7. Handicap usually begins when the average hearing loss for 500, 1000, and 2000 hertz is 25 dB.

13.4 ANECHOIC CHAMBER

Acoustical measurements under free-field conditions are required in the development of the major portion of electroacoustic transducers. The most obvious and direct solution would seem to be to make the measurements out-of-doors at a great distance from all reflecting surfaces. There are several objections to outdoor testing, for example, interruptions due to wind, rain, and snow; noise, both natural and man made; and difficulty in arranging experiments at a sufficient distance from the earth so that reflections will be negligible. In view of the importance of free-field testing and the objections to outdoor arrangements, it is obvious that a free-field sound room is an almost indispensable part of the equipment of an acoustical laboratory.

The term free-field room is used to designate a room in which free-field acoustic conditions are obtained, that is, a room in which the reflections from the boundaries are negligible. Such rooms have been termed anechoic chambers. The word anechoic is made up from the Greek prefix *an-*, meaning not or without, the Greek word *echo*, meaning an echo or a sound reflection, and the English adjectival suffix *-ic*, meaning characterized by.

The objective in the design of an anechoic chamber is to reduce to a negligible amount all reflections from the boundary surfaces of the room. This is equivalent to a very small ratio of generally reflected to direct sound. The ratio of the generally reflected to the direct sound in a room is as follows:

$$\frac{E_R}{E_D} = 16\pi D^2 \frac{(1 - a)}{aS} \tag{13.1}$$

where E_R = energy density of the reflected sound, in ergs per cubic centimeter,

E_D = energy density of the direct sound, in ergs per cubic centimeter,

D = distance from the source to the observation point, in centimeters,

S = area of the absorbing material, in square centimeters, and

a = absorption coefficient of the boundaries of the room.

An examination of Equation (13.1) shows that the ratio of reflected to direct sound may be reduced by decreasing the distance between the source and observation point, by making the absorption coefficient of the walls near unity, or by increasing the area of the walls. In other words, free-field conditions are approached by making the room large and the absorption coefficient of the wall near unity.

Three types of wall treatment for anechoic chambers are shown in Fig. 13.8. Fig. 13.8X depicts a baffle type of sound treatment. In the preferred design the spacing between the baffles is 6 inches. Each baffle is made up of uncompacted Ozite 2 inches in thickness. Parts Y and Z of Fig. 13.8 show sound treatment in the form of pyramids and wedges made of fiber glass. The general consensus is that the wall treatment of Fig. 13.8Z exhibits the greatest sound-absorbing efficiency. In this connection, it should be mentioned that sound-absorbing efficiency beyond a certain point is of little practical significance under actual operating conditions, when reflecting surfaces almost invariably are introduced into the room in almost any measurement.

The portion of the frequency range where it is difficult to provide free-field conditions is the low frequency range. In this connection, regardless of the type of wall treatment employing passive materials, the sound absorption deviates rapidly from unity when the depth of the material is less than one-quarter of a wavelength of the sound. Furthermore, free-field conditions can only be obtained when the dimensions of the room are each greater than one wavelength of the sound.

An anechoic chamber is shown in Fig. 13.9. The walls, ceiling, and floor are all covered with a wall treatment of the square-base pyramid type. A sound-transparent grid is used to support the apparatus being tested. The

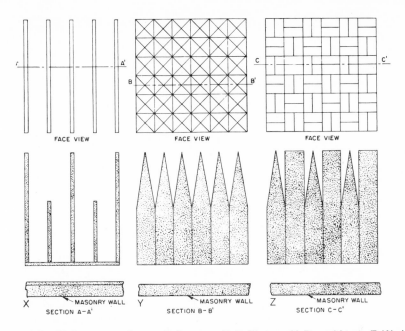

Fig. 13.8 Wall treatment for anechoic rooms. X. Baffle type. Y. Pyramid type. Z. Wedge type.

grid is located so that the measurements are made in the central volume of the room. The outside walls, ceiling, and floor are of heavy masonry so that extraneous noises are excluded. A noise level of near 0 dB is a desired objective for an anechoic chamber.

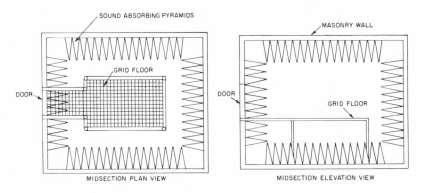

Fig. 13.9 Sectional views of an anechoic chamber.

13.5 AUDIO FREQUENCY RESPONSE RECORDER

An audio frequency response recorder consists of an audio oscillator and automatic level recorder in which the paper record drive is mechanically connected to the means for varying the frequency of the oscillator as shown in Fig. 13.10. The audio frequency response recorder may be used to measure the frequency response of microphones, loudspeakers, earphones, amplifiers, reverberation time, etc. Some of the measurements which can be made by the audio frequency response recorder will be described in the sections which follow.

13.6 LOUDSPEAKER FREQUENCY RESPONSE MEASUREMENTS

The arrangement of the apparatus for obtaining the free-field frequency response characteristic of a loudspeaker is shown in Fig. 13.10. The loudspeaker and calibrated microphone are set up in the anechoic room as shown in the figure. The audio frequency response recorder is compensated for any response variations of the microphone with respect to frequency so that the response obtained on the loudspeaker represents the true characteristic.

The total sound power output of a loudspeaker may be obtained by the procedure outlined in Section 2.6. Frequency response measurements are made as depicted in Fig. 13.10 at the angles required to carry out the integrating procedure of Equation (2.22) and Fig. 2.34.

Another method is to obtain the frequency response in a highly reverberant room with a reverberation time of more than 5 seconds, the idea for so doing being that all the sound which strikes the microphone is reflected sound and therefore represents the sound power output of the loudspeaker.

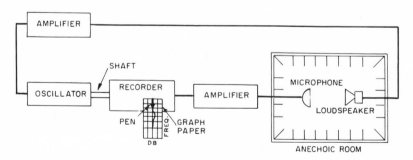

Fig. 13.10 Schematic diagram of the elements of the equipment for obtaining the frequency response of a loudspeaker in an anechoic room.

13.7 RECIPROCITY CALIBRATION OF MICROPHONES

The three experiments for the free-field reciprocity calibration of a microphone are shown in Fig. 13.11. For the application of the reciprocity principle to the free-field calibration of a microphone the following three transducers are used: the microphone M, to be calibrated, a reversible microphone-loudspeaker S_1, and a loudspeaker S_2. For the reversible microphone-loudspeaker it is convenient to use a small back-enclosed loudspeaker.

The absolute response of the microphone is given by:

$$K = \sqrt{\frac{2d\lambda e_M e'_M}{r_A i e_S}} \qquad (13.2)$$

where K = open-circuit voltage of the microphone, in abvolts per microbar,

d = distance between transducers in all of the experiments, in centimeters,

λ = wavelength of the sound, in centimeters,

$r_A = \rho c$,

ρ = density of air, in grams per cubic centimeter,

c = velocity of sound, in centimeters per second,

e_S = open-circuit output of the microphone-loudspeaker S_1 in experiment A, in abvolts,

e_M = open-circuit output of the microphone M in experiment B, in abvolts,

e'_M = open-circuit output of the microphone M in experiment C, in abvolts, and

i = current input to the microphone-loudspeaker in experiment C, in abamperes.

In order to produce the same sound pressure at the microphone-loudspeaker S_1 in experiment A and the microphone M in experiment B, the input to the loudspeaker S_2 must be the same in both experiments A and B. Furthermore, the distance d must be the same for both experiments A and B.

As indicated earlier in the text, the CGS Electromagnetic System of units is used for the electrodynamic microphones, as exemplified by the dynamic, ribbon, and magnetic microphones. For the crystal and condenser microphones the CGS Electrostatic System of units is used.

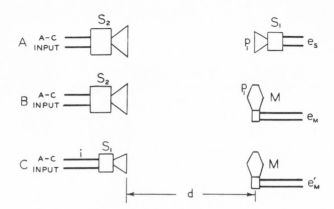

Fig. 13.11 The three experiments of the reciprocity procedure for obtaining the free field calibration of a microphone. A. The open circuit voltage, e_S, of the reversible microphone-loudspeaker S_1, when used as a microphone and actuated by a sound pressure p_1. B. The open circuit voltage, e_M, of the microphone, M, to be calibrated when actuated by a sound pressure p_1. C. The open circuit voltage e'_M, of the microphone, M, to be calibrated when actuated by a sound pressure produced by the reversible microphone-loudspeaker, S_1, used as a loudspeaker with a current input i, and a separation d.

13.8 EARPHONE FREQUENCY RESPONSE MEASUREMENTS

The arrangement of the apparatus for obtaining the frequency response characteristic of an earphone is shown in Fig. 13.12. The earphone is driven by the amplified output of the audio oscillator. The artificial ear consists of a condenser microphone coupled to a cavity of 4 cubic centimeters. The earphone is placed over the cavity, the output of the microphone is amplified and fed to the recorder, and the response frequency characteristic of the earphone is obtained by means of the audio frequency response recorder, as depicted in Fig. 13.12.

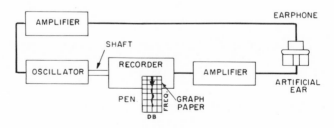

Fig. 13.12 Schematic diagram of the elements of the equipment for obtaining the frequency response of an earphone employing an artificial ear.

13.9 NONLINEAR DISTORTION MEASUREMENTS

The arrangement of the apparatus for measuring the nonlinear distortion of a loudspeaker is shown in Fig. 13.13. In general, the nonlinear distortion of a microphone for the levels encountered in sound reproduction is negligible. However, this is not the case for the loudspeaker, particularly in the low frequency range. The nonlinear distortion meter consists of a very-narrow-frequency band, variable frequency filter, and an electronic voltmeter. The response of the narrow-frequency filter is shown in Fig. 13.13. The narrow-frequency band of the filter makes it possible to measure the harmonic nonlinear distortion over the entire audio frequency band. The total nonlinear harmonic distortion in percent is the ratio of the sum of the harmonics to the fundamental multiplied by 100.

The apparatus of Fig. 13.13 may also be used to measure the harmonic nonlinear distortion of other elements of a sound-reproducing system.

13.10 TRANSIENT RESPONSE MEASUREMENTS

A measure of the transient response of a loudspeaker or other audio equipment may be obtained by measuring the response to an electrical input in the form of a tone burst. The term "tone burst" is used to designate a series of sine wave cycles with a rectangular envelope. A schematic diagram of the apparatus for obtaining the transient response of a loudspeaker is shown in Fig. 13.14. An audio oscillator supplies the sine wave signal, which is fed to an electronic gate that interrupts the sine wave signal at regular intervals, thereby supplying the tone burst. The tone bursts are amplified and fed to the loudspeaker. The sound output from the loudspeaker is picked up by the microphone. The output of the microphone is amplified and applied to the vertical deflection system of a cathode-ray oscilloscope. The acoustical output of the loudspeaker may be compared to the electrical input to the

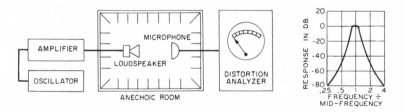

Fig. 13.13 Schematic diagram of the elements of a variable narrow-frequency band harmonic analyzer for obtaining the harmonic nonlinear distortion of a loudspeaker. The graph depicts the frequency response of the narrow-band filter.

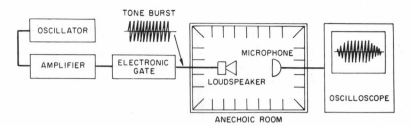

Fig. 13.14 Schematic diagram of the elements of the equipment used in the tone burst system for depicting the transient response of a loudspeaker.

loudspeaker. The deviation between the input and output tone bursts depicts the transient response distortion.

The above experiment assumes that transient response of the microphone is free of transient distortion. In general, microphones with simple vibrating systems and smooth frequency response characteristics, such as the ribbon velocity and condenser microphones, exhibit excellent transient response. A loudspeaker with a smooth frequency response characteristic will usually exhibit good transient response.

The transient response of amplifiers, earphones, and other audio equipment may also be obtained by the use of the tone burst.

13.11 REVERBERATION-TIME MEASUREMENTS

The reverberation time of an enclosure for a given frequency is the time required for the sound energy density, initially in a steady state, to decrease after the sound is stopped to one-millionth of the initial or steady state sound energy density. The unit is the second.

A schematic diagram of the apparatus for obtaining the reverberation time of a room is shown in Fig. 13.15. The output of the oscillator is amplified and fed to the loudspeaker. The sound level of the sound energy density is measured by means of the microphone, amplifier, and audio frequency level recorder. When the steady state conditions obtain, the input signal to the loudspeaker is stopped and the decay of the sound level is recorded by the audio frequency level recorders. The reverberation time is determined from the speed of the paper and amplitude decay.

13.12 LOUDNESS MEASUREMENTS

The sounds of speech, music, and noise present an infinite variety of numberless combinations of frequencies and intensities with respect to time. In order to provide a measure of the loudness for the complex sounds of speech, music, and noise, there must be some means for separating the complex sounds into

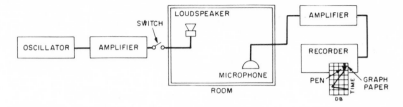

Fig. 13.15 Schematic diagram of the elements of the equipment for obtaining the reverberation time of a room.

manageable segments. In particular, to establish the loudness of a complex sound at least three specifications must be available as follows:

(1) A scale of subjective loudness, which is termed the sone (the relation between sones and phons is given by Equation (17.1).

(2) The equal loudness contours for discrete frequency bands of the complex sound.

(3) The rule by which loudness adds as the discrete frequency bands of the complex sound are added.

Specifically the sound pressure level in each octave band is determined by means of the sound level meter of Fig. 13.3. When the sound pressure level in each octave band has been determined, the next step is the proper summation of these data to provide the total or overall loudness of the complex sound. In accordance with the ISO-R532 Method A Standard, the relation between the total loudness and the loudness index is given by:

$$S_T = 0.7S_M + 0.3 \sum S \qquad (13.3)$$

where S_T = total loudness of the complex sound, in sones,
S = loudness index in each octave band, and
S_M = greatest of the loudness indices.

The loudness index in each octave band is obtained from the graph of Fig. 13.16.

To illustrate this procedure, the following calculation was carried out on the noise in a machine shop.

Octave Band in Hz.	Band Level dB	Band Loudness Index
63	68	2.8
125	69	4.7
250	70	6.2
500	70	7.4
1,000	71	9.3
2,000	73	12.6
4,000	69	11.8
8,000	64	10.5

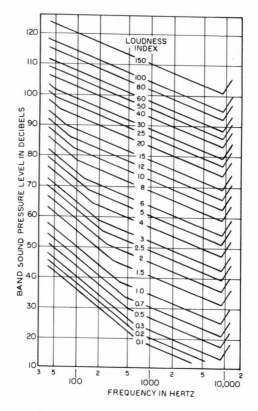

Fig. 13.16 Contours of the equal loudness index for octave bands in the audio frequency range.

Using the data in the table of calculations:

$$\sum S = \text{sum of the band loudness indices} = 65.3$$
$$S_M = \text{maximum band loudness index} = 12.6$$
$$0.3 \sum S = 19.6$$
$$0.7S_M = \underline{\quad 8.8}$$
$$\text{loudness (in sones)} = 0.3 \sum S + 0.7S_M = 28.4 \text{ sones}$$

From Equation (17.1), 28.4 sones = 88.2 phons.

A schematic diagram of a loudness meter based upon the above procedure is shown in Fig. 13.17. The system consists of octave band filters, phon-to-sone converters, differential gates for selecting the highest sone level, a means for adding the total sone output, attenuators which satisfy Equation (13.3), and a meter which indicates the loudness in sones. The loudness meter shown in Fig. 13.17 provides an instrument for the real time measurement of loudness.

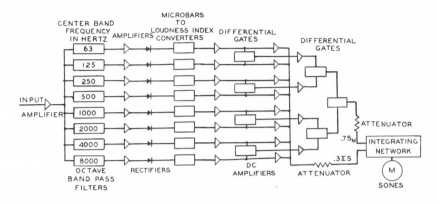

Fig. 13.17   A schematic diagram of the elements in a loudness meter.

13.13 PERCEIVED NOISE LEVEL MEASUREMENTS

The measurement of loudness has been described in the preceding section. There appears to be a need for the measurement of perceived noise, that is, the "noiseness" of a noise. The relation between the perceived noise or noiseness, in noys, and the noise index is given by:

$$N_T = 0.7N_M + 0.3 \sum N \tag{13.4}$$

where N_T is the total perceived noise of a complex sound, in noys; N is the noys index, in each octave band; and N_M is the greatest of the noys indices. The noys index is obtained from the graph of Fig. 13.18.

The calculations employing Equation (13.4) are carried out in the same manner as are those in the case of loudness. When the total perceived noise is obtained in noys, the noys are converted into decibels above threshold by using the 1000-hertz relation between noys and sound pressure level of Fig. 13.18.

Referring to Figs. 13.16 and 13.18, there is not a great deal of difference between loudness and noys.

13.14 ARTIFICIAL VOICE

The performance of microphones under actual operating conditions may be determined by means of an artificial voice. There are two types of artificial-voice system, namely, the field type, in which there is a considerable distance between the microphone and the artificial voice, and the close-talking type, in which the distance between the microphone and the artificial voice is the matter of inches.

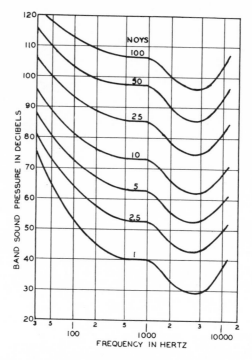

Fig. 13.18 Contours of equal perceived noise (noys) for octave bands in the audio frequency range.

A. Field-Type Artificial Voice

The field-type artificial voice consists of an enclosure with the dimensions of the average human head that houses a small loudspeaker system feeding through a small opening as depicted in Fig. 13.19. The loudspeaker is designed to provide uniform response over the frequency range of 80 to 12,000 hertz. The enclosure is designed to provide the same directivity pattern as the human voice. The artificial voice should be capable of delivering a level of 80 dB at a distance of 5 feet. The microphone should not be any closer than 6 inches from the mouth of the field-type artificial voice.

To use the field-type artificial voice, a recording of a live voice is carried out in an anechoic chamber as shown in Fig. 13.19A. The recording is played through a magnetic tape recorder, the output of which is connected to the artificial voice. Under these conditions the output of the artificial voice is a replica of the original voice in all respects.

In Fig. 13.19B tests are carried out on floor-stand and lavalier microphones.

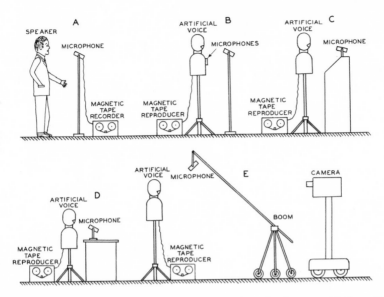

Fig. 13.19 Field type artificial voice. A. Recording the voice for use with the artificial voice. Testing microphones under different conditions are depicted as follows; B. Floor stand and lavalier. C. Lecturn mounted. D. Desk stand. E. Boom mounted.

In Fig. 13.19C tests are carried out on a lecturn microphone. In Fig. 13.19D tests are carried out on a desk-stand microphone. In Fig. 13.19E tests are carried out on a boom-mounted microphone.

Several types of test may be employed. The performance of a sound system may be determined by use of the arrangements of Fig. 13.19B, C, and D. The speech pickup in a sound motion picture or television system may be determined by all the arrangements depicted in Fig. 13.19B, C, D, and E. Continuous speech may be employed and the listening carried out at the output points, which may be various locations in an auditorium or monitoring room in a sound motion picture or television studio. Syllables and words may be used to obtain the syllable articulation and word intelligibility of the system.

The advantage of the artificial voice over that of a live speaker is that the system may be operated over prolonged periods of time with no changes in the voice. In the case of the live voice, its quality changes when the voice becomes tired after long periods of talking.

Frequency response characteristics may also be obtained, by means of an oscillator and amplifier driving the artificial voice and recording the output of the system.

B. Close-Talking Artificial Voice

A close-talking artificial voice consists of a small loudspeaker driving an acoustic line housed in an enclosure simulating the human head, as depicted in Fig. 13.20. The acoustic impedance of the opening of the acoustic line simulates the acoustic impedance of the human voice. Under these conditions the reaction of the artificial voice upon a microphone placed close to the mouth will be the same as for the human voice. The frequency response of a microphone using the artificial voice may be obtained in a manner similar to that in Fig. 13.10 except that the artificial voice is substituted for the loudspeaker.

Subjective tests may also be carried out by means of the close-talking artificial voice in a manner similar to that described for the field-type artificial voice described in the preceding section.

The particular design of artificial voice shown in Fig. 13.20 illustrates the general concept. There are actually many different designs, all of which satisfy the main requirement of a "close-talking" artificial voice, namely, that the reaction upon the microphone be the same as that of the human voice.

13.15 DISK PHONOGRAPH MEASUREMENTS

The frequency response performance of a disk phonograph pickup can be obtained by means of a calibrated frequency disk record, which is commercially available.

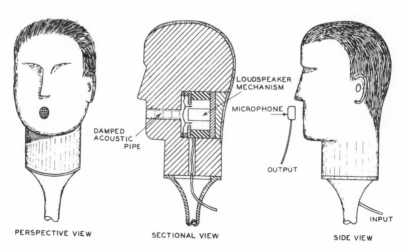

Fig. 13.20 Perspective, sectional, and side views of a close talking artificial voice. The side view shows a microphone mounted in close proximity to the mouth.

13.16 MAGNETIC TAPE RECORDER AND REPRODUCER MEASUREMENTS

The overall frequency response of a magnetic tape recorder and reproducer can be obtained by means of the oscillator recorder system depicted in Fig. 13.10 in which the recorder and reproducer are substituted for the loudspeaker and microphone.

13.17 VOLUME INDICATOR

A volume indicator is an instrument used to measure the power level of audio signals. The most common volume indicator is the VU (volume unit) meter, consisting of a 200-microampere direct current D'Arsonval movement fed from a copper oxide full-wave rectifier. The scale of the meter covers the range -20 to $+3$ VU. The indicated reading is generally not the absolute value but is related by the attenuator setting or gain change to some desired reference level, for example $+4$VU in a studio line. A reading of 0 VU corresponds to 1 milliwatt of power in a 600-ohm circuit.

REFERENCES

Audiometers for General Diagnostic Purposes, American Standards Association, ASA, 24.5-1951.

Bauer, B. B. and Torick, E. L. "Researches in Loudness Measurement," *IEEE Transactions on Audio and Electroacoustics*, Vol. AU-14, No. 3, p. 141, 1966.

Bauer, B. B., Torick, E. L., Rosenheck, A. J., and Allen, R. G. "A Loudness Level Monitor for Broadcasting," *IEEE Transactions on Audio and Electroacoustics*, Vol. AU-15, No. 4, p. 177, 1967.

Brock, Jens Trampe. *Acoustic Noise Measurement*, Bruel and Kjaer, Cleveland, Ohio, 1969.

Chinn, H. A., Gannett, D. K., and Morris, R. M. "A New Standard Volume Indicator," *Proceedings of the Institute of Radio Engineers*, Vol. 28, No. 1, p. 1, 1940.

Davis, Don. *Acoustical Tests and Measurements*, Howard W. Sams, Indianapolis, 1965.

Howard, Richard. "An Improved Audiometric Earphone," *Journal of the Audio Engineering Society*, Vol. 16, No. 3, p. 328, 1968.

Keast, David. *Measurements in Mechanical Dynamics*, McGraw-Hill, New York, 1967.

Kryter, K. D. and Pearsons, K. S. "Some Effects of Spectral Content and Duration on Perceived Noise Level," *Journal of the Acoustical Society of America*, Vol. 35, No. 6, p. 866, 1963.

Olson, Harry F. *Acoustical Engineering*, Van Nostrand Reinhold, New York, 1957.

Olson, Harry F. "Artificial Voice," Audio Engineering Society, Preprint No. 389, Convention, October 15, 1965.

Peterson, Arnold P. G. and Gross, Ervin E. Jr. *Handbook of Noise Measurement*, General Radio Company, West Concord, Mass., 1967.

"Precision Sound Level Meters," International Electrotechnical Commission, IEC/179, 1965.

Schwatz, A., Gust, A. J., and Bauer, B. B. "Absolute Calibration of Pickup and Records," *Journal of the Audio Engineering Society*, Vol. 14, No. 3, p. 218, 1966.

Specifications for Audiometers, American National Standards Institute, ANSI, s3.6-1969.

Standard Reference Zero for Calibration of Pure Tone Audiometers, International Organization for Standardization, ISO Recommendation 389-1964.

Stevens, S. S. "Procedure for Calculating Loudness, Mark IV," *Journal of the Acoustical Society of America*, Vol. 33, No. 11, p. 1577, 1961.

Torick, E. L., Di Mattia, A., Rosenbeck, A. J., Abbaagnaro, I. A., and Bauer, B. B. "An Electronic Dummy for Acoustical Testing," *Journal of the Audio Engineering Society*, Vol. 16, No. 4, p. 397, 1968.

14
Acoustical
Performance of
Enclosures

14.1 INTRODUCTION

The acoustical performance of a sound-collecting or dispersing system in an enclosure is exceedingly complex because so many parameters are involved, as for example, the cubical content, the shape and the sound absorption of the enclosure, the ambient noise, and all the characteristics of the sound pickup or sound-reproducing equipment. The purpose of this chapter is to provide an exposition on the collection and dispersion of sound in an enclosure involving the interplay of sound reproduction electroacoustics and room acoustics.

14.2 DISPERSION OF SOUND

The dispersion of sound in a room involves the performance of a sound source in a room, whether it be an original sound source, as for example, the voice, a musical instrument, and so on, or a reproducing sound source, as for example, a loudspeaker. The purpose of this section is to provide an exposition on the dispersion of sound in a room by a sound source.

A. Nondirectional Sound Source Operating in an Enclosure

When a source of sound is started in a room, the energy does not build up instantly, a fact that is due to the finite velocity of a sound wave. Each pencil of sound sent out by the sound source is reflected many times by the sound-absorbing walls of the room and thus ultimately dissipated. A nondirectional source of sound operating in a room is shown in Fig. 14.1A. The equation for the direct sound energy at the observation point is given by:

$$E_D = \frac{P_O}{4\pi d^2 c} \tag{14.1}$$

where E_D = direct sound energy density due to the direct sound, in ergs per cubic foot,

P_O = power output of the sound source, in ergs per second;

d = distance between the sound source and the observation point, in feet, and

c = velocity of sound, in feet per second.

The direct sound energy density is depicted in Fig. 14.2.

The equation for the growth of the sound energy density in a room with a nondirectional sound source operating is given by:

$$E_{RN} = \frac{4P_O}{caS} (1 - \varepsilon^{cS\,[\log_e (1-a)]t/4V})(1 - a) \tag{14.2}$$

where E_{RN} = reflected sound energy density, in ergs per cubic foot, at a time t seconds after starting the sound source,

P_O = power output of the sound source, in ergs per second,

c = velocity of sound, in feet per second,

S = area of the boundaries of the room, in square feet,

a = average sound absorption per square foot, in sabins, and

V = volume of the room, in cubic feet.

The sound energy density growth characteristic of the sound in a room obtained from Equation (14.2) is shown in Fig. 14.2.

A steady state condition obtains in Fig. 14.1A when the energy absorbed by the walls equals the energy emitted by the sound source. From Equation (14.2) the sound energy density, when a steady state condition obtains, is given by:

$$E_{RN} = \frac{4P_O}{caS} (1 - a) \tag{14.3}$$

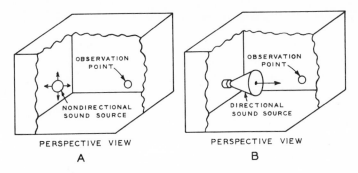

Fig. 14.1 Perspective views of a room containing a sound source and an observation point. A. A nondirectional sound source. B. A directional sound source.

The sound energy density steady state characteristic of the sound in a room, obtained from Equation (14.3), is shown in Fig. 14.2.

The equation for the decay of the sound energy density in the room of Fig. 14.1 is given by:

$$E_{RN} = \frac{4P_O}{caS} \left(\varepsilon^{cS\,[\log_e (1-a)]t/4V} \right) (1 - a) \qquad (14.4)$$

The sound energy density decay characteristic of the sound in a room obtained from Equation (14.4) is shown in Fig. 14.2.

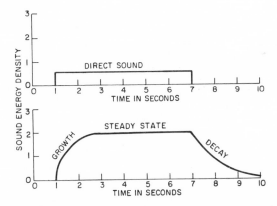

Fig. 14.2 A typical illustration of the sound energy densities of the direct and reflected sounds in a room. The reflected sound shows the growth, steady state and decay conditions.

B. Reverberation Time of an Enclosure

The reverberation time of a room for a sound of a given frequency is the time required for the average sound energy density, initially in a steady state condition, to decrease, after the cessation of the sound source, to one-millionth of the initial value. The unit is the second.

From Equation (14.4), the time required for sound energy density to decay to one-millionth of the original intensity is given by:

$$t_R = \frac{0.05V}{-S \log_e (1 - a)} \quad \text{or} \quad \frac{0.05V}{-2.3S \log_{10} (1 - a)} \tag{14.5}$$

where t_R = reverberation time, in seconds,
  V = volume, in cubic feet,
  S = total area of the walls, ceiling, and floor, in square feet, and
  a = average absorption coefficient of the surfaces enclosing the room.

C. Absorption Coefficient

The acoustic absorption coefficient of a surface is the ratio of the flow of sound energy into the surface on the side of incidence to the incident rate of energy flow. The sabin is the unit of equivalent absorption and is equal to the equivalent absorption of 1 square foot of a surface of unit absorptivity, that is, 1 square foot of surface which absorbs all incident sound energy.

A tabulation of sound absorption coefficients for acoustical materials, building materials, and individual objects is shown in Table 14.1.

D. Sound Transmission Through Partitions

The isolation of a sound source that involves the transmission of sound through partitions is a subject closely allied to the absorption of sound in a room. Furthermore, the isolation of a room from the sounds of an adjacent room or from outside noise, which would interfere with the pickup of sound or the reproduction of sound, is of importance in the acoustical performance of enclosures.

The problem of sound transmission through partitions and walls is complicated because of the many factors involved. The performance of the mass-controlled single-wall partition is very simple. The sound insulation of this type of partition is proportional to the mass and frequency. For the usual building materials and walls of ordinary dimensions supported at the edges, the problem is that of the clamped rectangular plate with distributed resistance throughout the plate and lumped damping at the edges. Obviously, the

TABLE 14.1

Sound Absorption Coefficients of Acoustical Materials, Building Materials, and Objects

Material	Thickness (in.)	Frequency					
		128	256	512	1024	2048	4096
		Coefficient					
Typical commercial acoustical materials	1	0.28	0.45	0.80	0.90	0.85	0.85
Draperies, 10 oz. per sq. yd., hung 4″ from the wall		0.09	0.33	0.45	0.52	0.50	0.44
Hair felt in contact with wall	1	0.13	0.41	0.56	0.69	0.65	0.49
Carpet, on 0.125″ felt, on concrete	0.4	0.11	0.14	0.37	0.43	0.57	0.67
Wood sheeting, pine	0.8	0.10	0.11	0.10	0.08	0.08	0.11
Concrete, unpainted		0.01	0.012	0.016	0.019	0.023	0.035
Brick wall, unpainted		0.024	0.025	0.031	0.042	0.049	0.070
Brick wall, painted		0.012	0.013	0.017	0.020	0.023	0.025
Concrete porus blocks, unpainted		0.15	0.21	0.43	0.37	0.39	0.51
Plaster, lime, wood lath on wood studs, rough finish		0.023	0.039	0.039	0.052	0.037	0.035
Individual Object		Absorption in Sabins					
Audience per person, man with coat		2.3	3.2	4.8	6.2	7.6	7.0
Auditorium chairs, solid hard seat and back		0.15	0.22	0.25	0.28	0.50	0.45
Auditorium chairs, upholstered		2.7	3.1	3.0	3.2	3.4	3.2

The above data represents typical performance. As in all acoustical applications there may be wide variations in the performance due to shape, size and other parameters of the enclosure. The sound absorption coefficients of typical commercial acoustical materials one inch in thickness is shown above. For sound absorption coefficients from different manufacturers and of various thickness the reader is referred to the Official Bulletin of the Acoustical and Insulating Materials Association (AIMA).

performance of this system depends upon the size, the ratio of the two linear dimensions, the weight of the material, the damping in the material, and the edge supports. This type of problem is not amenable to an analytical solution.

The transmittivity of a partition is defined as the ratio of the intensity in the sound transmitted by the partition to the intensity in the sound incident upon the partition. The transmission loss, in decibels, introduced by the partition is given by:

$$TL = 10 \log_{10} \frac{I_i}{I_t} = 10 \log_{10} \frac{1}{\tau} \tag{14.6}$$

where TL = transmission loss, in decibels,
    I_i = intensity of the incident sound,
    I_t = intensity of the transmitted sound, and
    τ = transmittivity or the transmission coefficient.

The coefficient of transmission τ is a quantity which pertains alone to the partition and is independent of the acoustic properties of the rooms which it separates.

The noise-reduction factor is the ratio of the sound energy density in the room containing the sound source to the sound energy in the adjoining receiving room. The noise-reduction factor, in decibels, is given by:

$$RF = TL + 10 \log_{10} \frac{A}{S} \tag{14.7}$$

where RF = noise-reduction factor, in decibels,
    TL = transmission loss, in decibels,
    A = total absorption in the receiving room, in sabins, and
    S = area of the partition, in square feet.

Equation (14.7) shows that the reduction is due to both the loss introduced by the partition and the absorption in the receiving room.

The choice of a partition for insulating a room against sound involves a number of considerations. Some of the factors are the frequency distribution and intensity level of the components of the objectionable sound, the transmission frequency characteristics of the partition, the ambient noise or sound level in the receiving room which will mask the objectionable sound, and the response frequency characteristics of the ear.

A tabulation of the transmission loss and the coefficient of transmission τ for partitions of various types is shown in Table 14.2.

The mass-controlled partition with air between the partition elements is a low-pass filter in which the mass of the wall is the series element and the volume between the partition is the shunt element. The partitions in this case are mounted in edge supports that allow freedom of motion without cracks which would pass air-borne sound.

TABLE 14.2

Sound Transmission Loss Through Various Partitions

Material	Weight in lbs./ft^2	Average Transmission loss in dB	Transmission Coefficient τ
Aluminum 0.025″	0.35	16	0.025
Iron 0.03″	1.2	25	0.0032
Lead 0.125″	8.2	32	0.00063
Plywood 0.25″	0.73	21	0.008
Typical acoustical material 1″	1.2	25	0.0032
Hair felt 1″	0.75	6	0.25
Wood studs 2″ × 4″ metal lath, plaster 0.75″	21	40	0.0001
Tile 2″ gypsum	20	42	0.000063
Cinder block 4″ plastered both sides	37	40	0.0001
Tile clay 6″ plastered both sides	37	40	0.0001
Brick 8″ plastered both sides	87	50	0.00001
Brick 16″ plastered both sides	174	55	0.0000032
Double cinder block, each 8″, air space 4″, outside plastered	132	60	0.000001
Door, light 4-panel		22	0.0063
Door, oak		25	0.0032
Door, steel 0.25″		35	0.00032
Window glass, plate 0.25″	3.5	30	0.001
Double window glass 0.25″, air space 1″	7.0	45	0.000032
Finish and rough flooring, plaster ceiling		37	0.0002
Concrete slab 4″	48	42	0.000063
Concrete slab 4″, suspended plaster ceiling	51	52	0.0000063

The above data represents typical performance. As in all acoustical applications there may be wide variations due to shape, size, and other parameters of the partitions and the enclosures.

E. Directional Sound Source Operating in an Enclosure

The performance of a nondirectional sound source operating in a room was considered in Section 14.2B. Since most sound sources are directional, in particular loudspeakers, the purpose of this section is to develop the equations for a directional sound source operating in a room.

The equation for the growth of the sound energy density in a room with a directional sound source operating is given by:

$$E_{RD} = \frac{P_o\Omega}{caS\pi} (1 - \varepsilon^{cS\,[\log_\varepsilon\,(1-a)]t/4V})(1 - a) \qquad (14.8)$$

where E_{RD} = reflected sound energy density, in ergs per cubic foot, at a time t seconds after starting the sound source, and

Ω = effective solid angle of the sound-dispersing source, in steradians.

All other quantities are the same as in Equation (14.2).

The total sound power output of the directional sound source is given by:

$$P_o' = P_o \frac{\Omega}{4\pi} \qquad (14.9)$$

where P_o' = power output of the sound source, in ergs per second.

From Equation (14.8), the sound energy density when a steady state condition obtains is given by:

$$E_{RD} = \frac{P_o\Omega}{caS\pi} (1 - a) \qquad (14.10)$$

The equation for the sound energy density for the decay of sound in the room of Fig. 14.1B is given by:

$$E_{RD} = \frac{P_o\Omega}{caS\pi} (\varepsilon^{cS\,[\log_\varepsilon\,(1-a)]t/4V})(1 - a) \qquad (14.11)$$

The implicit assumption in Equations (14.1), (14.2), (14.3), (14.4), (14.8), (14.9), (14.10), and (14.11) is that the same direct sound energy density exists at the observation points in both A and B of Fig. 14.1. The direct sound energy density of the direct sound at the observation point in Fig. 14.1A or B is given by:

$$E_D = \frac{P_o}{4\pi d^2 c} = \frac{P_o'}{d^2 c\Omega} \qquad (14.12)$$

where E_D = sound energy density due to the direct sound, in ergs per cubic foot, and

d = distance between the sound source and the observation point, in feet.

F. Effective Reverberation of Reproduced Sound

The preceding sections show that in the case of a loudspeaker operating in a room as depicted in Fig. 14.1B there are two sources of sound at the observation point, namely, the direct and the reflected sound. The effective reverberation of reproduced sound is the ratio of the reflected to the direct sound energy density at the observation point.

The ratio of the reflected to the direct sound energy density at the observation point in Fig. 14.1B from Equations (14.8) and (14.12) is as follows:

$$\frac{E_{RD}}{E_D} = \frac{4d^2\Omega}{aS} (1 - \varepsilon^{cS\,[\log_\varepsilon(1-a)]t/4V})(1 - a) \qquad (14.13)$$

When steady state conditions obtain, the ratio of the reflected to the direct sound energy density at the observation point is given by:

$$\frac{E_{RD}}{E_D} = \frac{4d^2\Omega}{aS} (1 - a) \qquad (14.14)$$

Equations (14.13) and (14.14) show that the effective reproduced reverberation at the observation point of Fig. 14.1B can be reduced by decreasing the distance d, by increasing the total sound absorption aS, or by decreasing the solid angle Ω of dispersion by the sound source. These facts are of importance in the performance of a loudspeaker operating in a room. (See Chapter 16.)

The above observations can be illustrated graphically, as shown in Fig. 14.3. The direct sound energy density at the observation points in Fig. 14.1A and B for a series of sounds of different durations is shown in Fig. 14.3A. The growth, steady state, and decay characteristics of the reflected sound energy density in a room with a long reverberation time for a nondirectional sound source is depicted in Fig. 14.3B and for directional sound source is depicted in Fig. 14.3C. The growth, steady state, and decay characteristics in a room with a short reverberation time for a nondirectional sound source is depicted in Fig. 14.3D and for a directional sound source is depicted in Fig. 14.3E.

The graphical illustration of Fig. 14.3 shows that the effective reverberation of the reproduced sound can be reduced by reducing the reverberation time or by the use of a directional sound source or by both.

When the reflected sound energy density is several times the direct sound energy density as in the case of Fig. 14.3B, the intelligibility of reproduced speech will be impaired. Furthermore, musical sounds will be blurred when there is a large ratio of reflected to direct sound coupled with considerable overlap of the reflected sounds, as in the case of Fig. 14.3B. Employing a directional sound source in Fig. 14.3C reduces the reflected sound energy

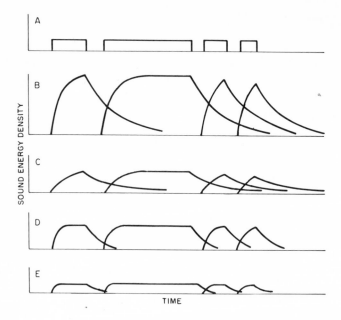

Fig. 14.3 Sound energy densities of the sounds in a room. A. Direct sound. B. Reflected sound in a room with a long reverberation time and a nondirectional sound source. C. Reflected sound in a room with a long reverberation time and a directional sound source. D. Reflected sound in a room with a short reverberation time and a nondirectional sound source. E. Reflected sound in a room with a short reverberation time and a directional sound source.

density to about two times that of the direct sound, an illustration of the effectiveness of a directional loudspeaker in reducing the effective reverberation of the reproduced sound. The reproduction of speech with practically perfect intelligibility will be obtained with the condition of Fig. 14.3E, in which the direct sound energy density is greater than the reflected sound energy density.

14.3 COLLECTION OF SOUND

The collection of sound in a room involves the performance of a sound-collecting system as exemplified by a microphone picking up the sound radiated by a sound source. The purpose of this section is to provide an exposition on the collection of sound in a room.

A. Sound-Collecting System Operating in an Enclosure

A sound-collecting system operating in a room is depicted in Fig. 14.4. The direct sound energy density available for actuating the microphone is given by:

$$E_D = \frac{P_O}{4\pi d^2 c} \qquad (14.15)$$

where E_D = direct sound energy density at the microphone, in ergs per cubic foot,

P_O = sound power output of the sound source, in ergs per second,

d = distance between the sound source and the microphone, in feet, and

c = velocity of sound, in feet per second.

Referring to Fig. 14.4, the effective response angle of the microphone is assumed to be Ω steradians. The direction and phase of the reflected sound

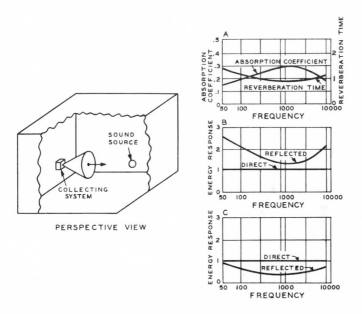

Fig. 14.4 Sound collecting system in a studio. A. The reverberation time of the studio and absorption coefficient of the boundaries of the studio. B. The energy response for the direct and reflected sounds for a nondirectional microphone. C. The energy response for the direct and reflected sounds for the bidirectional cosine velocity microphone or the cardioid unidirectional microphone.

are assumed to be random. Therefore, the reflected sound available for actuating the directional microphone consists of the pencils of sound within that angle. The energy response of the directional microphone to generally reflected sound will be $\Omega/4\pi$ times that of a nondirectional microphone. The generally reflected sound energy density to which the directional microphone responds is given by:

$$E_R = \frac{P_0\Omega}{caS\pi} (1 - \varepsilon^{cS\,[\log_\varepsilon\,(1-a)]t/4V})(1 - a) \qquad (14.16)$$

where E_R = reflected sound energy density capable of actuating the microphone, in ergs per cubic foot,

Ω = solid angle of reception by the microphone, in steradians,

a = average sound absorption per square foot, in sabins,

S = area of the boundaries of the room, in square feet,

V = volume of the room, in cubic feet, and

t = time after the sound source has started, in seconds.

The other quantities are the same as in Equation (14.15).

B. Effective Reverberation of the Collected Sound

The preceding discussion shows that there are two sources of sound which actuate the microphone, namely, the direct and the reflected sound. The ratio of the generally reflected to the direct sound energy density at the microphone is a measure of the effective reverberation of the collected sound.

The ratio of the generally reflected to the direct sound energy density capable of actuating the microphone from Equations (14.15) and (14.16) is as follows:

$$\frac{E_R}{E_D} = \frac{4d^2\Omega}{aS} (1 - \varepsilon^{cS\,[\log_\varepsilon\,(1-a)]t/4V})(1 - a) \qquad (14.17)$$

When steady state conditions obtain, the ratio of the reflected to direct sound energy density capable of actuating the microphone is given by:

$$\frac{E_R}{E_D} = \frac{4d^2\Omega}{aS} (1 - a) \qquad (14.18)$$

Equations (14.17) and (14.18) show that the effective collected reverberation can be reduced by decreasing the distance d, by increasing the total sound absorption, or by decreasing the solid angle Ω of reception by the microphone.

For a given room employing a directional microphone, the receiving distance d can be increased by $\sqrt{4\pi/\Omega}$ times that of a nondirectional microphone for the same collected reverberation in both cases.

The frequency absorption coefficient characteristic and reverberation time characteristic are shown in Fig. 14.4A. The direct sound picked up by a nondirectional microphone is shown in Fig. 14.4B and by a unidirectional microphone in Fig. 14.4C. The direct sound picked up by the microphone is the same because the distance between the sound source and the microphone is assumed to be the same for the two cases. The generally reflected sound picked up by a nondirectional microphone is shown in Fig. 14.4B. The generally reflected sound picked up by a cosine bidirectional velocity microphone or a cardioid unidirectional microphone, with in both cases $\Omega = 4\pi/3$, is shown in Fig. 14.4C. The effectiveness of a directional sound-collecting system in overcoming reverberation and undesirable sounds is graphically depicted in Fig. 14.4B and C.

REFERENCES

Bulletin of the Acoustical Materials Association, New York, 1970, "Sound Absorption of Materials."

Olson, Harry F. *Acoustical Engineering*, Van Nostrand Reinhold, New York, 1957.

Rettinger, Michael. *Acoustics, Room Design, and Noise Control*, Chemical Publishing Co., New York, 1968.

Weast, Robert C., Editor. *Handbook of Chemistry and Physics*, Chemical Rubber Company, Cleveland, 51st Edition, 1970–1971, "Sound Absorption Tables."

15
Sound
Collection in
Studios

15.1 INTRODUCTION

The collection of sound in an enclosure was considered in Chapter 14. The fundamentals of sound pickup in studios were developed in Section 14.3. Studios are employed for the pickup of sound for radio broadcasting, for recording of disk and magnetic tape records, for recording sound motion pictures, and for television broadcasts. The general arrangement, the equipment, and acoustics are all different for the four applications. The purpose of this chapter is to provide an exposition on studios and the sound pickup techniques for the different systems used in sound reproduction.

15.2 RADIO BROADCAST STUDIOS

Most of the material used for radio sound-broadcasting today is in the form of live speech for news and commentary, disk records, prerecorded magnetic tapes, and other specific prerecorded material. There are still a number of broadcasts for which the origin is a broadcast studio. The sound broadcast studio with a large audience no longer exists.

A broadcast studio with a setup for the sound pickup for a radio broadcast is shown in Fig. 15.1. The monitor room is located next to the studio. A large window between the studio and the monitor room makes it possible for the monitor engineer, the announcer, and the sound engineer to view the action in the studio. Communication between the monitor room and the studio is accomplished by a talk-back system as depicted in Fig. 15.1. Separate microphones are used for the actors and the announcer. The transcription turntable and magnetic tape reproducer supplies the audio signals from the recorded material. The program in the studio may also be recorded on magnetic tape for broadcast at a later time. In addition to musical and speech programs, a large segment of radio broadcasting consists of news, weather, and so on. The commentator for such broadcasts may be located in the studio or in the monitor room.

15.3 RECORDING STUDIOS

Recording studios are used for recording the master records for disk records, prerecorded magnetic tapes and cartridges, and musical elements of sound motion pictures.

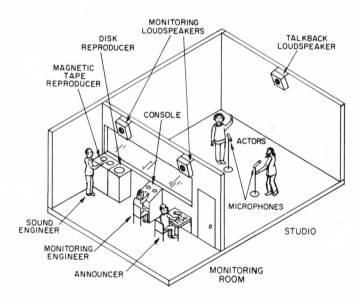

Fig. 15.1 Perspective view of a radio studio, monitoring room, equipment and personnel employed in radio broadcasting.

A recording studio for recording the master records for the above applications is shown in Fig. 15.2. The studio complex consists of the studio where the performers are located and the monitor room where the director, the monitoring engineer, and the recording engineer are located. The monitor room houses the console, the magnetic tape recorder, and sound-monitoring equipment. A large soundproof window between the monitoring room and the studio makes it possible for the director and the monitoring and recording engineers to view the action in the studio. Communication between the monitor room and the studio is accomplished by means of a talk-back system as shown in Fig. 15.2. The quality of the recorded sound is checked by the monitoring system in the monitoring room. Great care is exercised to provide high-quality sound-monitoring, to insure thereby a good recorded product.

The size of recording studios ranges from small studios in which only a few performers can be accommodated to studios that can be used to record a one-hundred-piece symphony orchestra.

In the early days of recording master records, two- and three-channel magnetic tape recorders were used. In the case of the three-channel recorder, the third channel was used for the vocalist or as an additional channel for

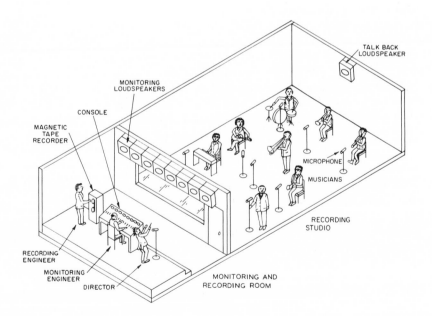

Fig. 15.2 Perspective view of the studio monitoring room, equipment and personnel employed in the recording of sound.

music. The microphones and musicians were arranged to provide the proper auditory perspective and balance among the sound sources.

The techniques for recording the master records for disk records, pre-recorded magnetic tapes, and the musical elements of sound motion pictures have undergone great changes in the past few years. The master recording process has employed more and more separate channels, the ultimate arrangement being one channel for each instrumentalist or vocalist. These techniques were evolved in the popular music field but now are being extended to semi-classical and classical fields.

In the example of Fig. 15.2 there is a separate microphone for each performer. In order to record each and every performer separately there must be no cross-talk between the various sound sources. To reduce the cross-talk either bidirectional or unidirectional microphones are used. In addition, a studio with a low reverberation time reduces the cross-talk. The separation between the microphone and the performer is made small and the separation between performers is made large to reduce further the acoustic cross-talk. If more isolation is required, baffles are placed between the performers. The output of each microphone is recorded on a separate track of a magnetic tape recorder. Multiple-channel magnetic tape recorders of 8, 16, and 24 tracks are available.

A schematic diagram of an eight-channel monitoring and recording system is shown in Fig. 15.3. Each sound source is monitored by means of the volume indicators and loudspeakers. The use of a loudspeaker in each channel makes it possible to detect acoustic cross-talk, noise, distortion, and other undesirable effects.

The eight-channel master magnetic tape record made on the system of Fig. 15.3 is modified and converted to a two-channel stereophonic record by means of the system shown in Fig. 15.4. The process is carried out in a room with dimensions and acoustics similar to that of the living room in a residence. When this means of monitoring is employed, the final product indicates the performance to be expected in the consumer's home

The system of Fig. 15.4 provides the means so that any of the eight original sound sources can be placed in either the left or right channel or any mixture in both channels. The important element in the system of Fig. 15.4 is the tone modifier. A schematic diagram depicting the elements of the tone modifier is shown in Fig. 15.5. There are sixteen such modifiers in the system of Fig. 15.4.

The delay in combination with the level makes it possible to specify the auditory perspective of any sound source, that is, place any sound source in reproduction at either the right or left loudspeaker or in any position between. The control of the auditory perspective is illustrated in Fig. 15.6. If the entire audio signal is fed to the left loudspeaker and no audio signal

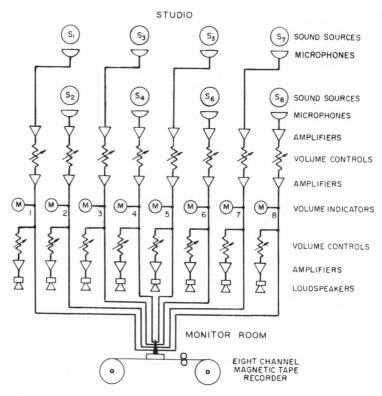

Fig. 15.3 The recording of eight sources each with a pickup microphone so that the signal outputs from the microphones are individually recorded on eight separate magnetic tracks on the tape.

to the right loudspeaker, the sound source will appear to be located at position 1. In the same way, if the entire audio signal is fed to the right loudspeaker and no audio signal to the left loudspeaker, the sound source will appear to be located at position 5. If the same audio signal is fed to both loudspeakers, the sound source will appear to be located at position 3. If delay is introduced in the left channel but the same amplitude is fed to both loudspeakers, the sound source can be moved from location 3 with no delay to location 5 with large delay to anywhere between 3 and 5, depending upon the amount of delay. If the delay in the two channels is the same but the signal in the left channel is attenuated, then the sound source can be moved from location 3 with no attenuation to location 5 with considerable attenuation to anywhere in between 3 and 5, depending upon the amount of attenuation. If both delay and attenuation are employed, the sound source can be placed

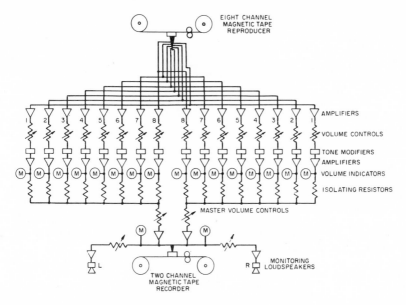

Fig. 15.4 The mixing, modifying and conversion to two channel stereophonic magnetic tape of the eight sound sources recorded in Figs. 15.2 and 15.3.

in any position at or between the loudspeakers. The attenuation and delay provides a powerful means for producing the desired auditory perspective.

Returning to the modifier of Fig. 15.5, the frequency response modifier consists of frequency selective networks that changes the original timbre by accentuating or attenuating frequency bands.

The timbre modifier of Fig. 15.5 is somewhat similar to the frequency response modifier except that new frequency components may be added by means of the timbre modifier.

The vibrato or tremolo generator modulates, by frequency or amplitude modulation or a combination of both, the original audio signal. The modulation frequency is of the order of 7 cycles per second.

Since the original sound as depicted in Figs. 15.2 and 15.3 is recorded without any reverberation, a means is provided to introduce artificial reverberation (see Section 12.6).

In general, the idea in all sound reproduction is to reduce nonlinear distortion. However, in some instances the subjective response can be heightened by the introduction of the appropriate nonlinear distortion. Accordingly, a means is provided for the introduction of nonlinear distortion.

The fuzz producer is a means of introducing high frequency components of large amplitudes and frequency-modified random noise. Here again, this

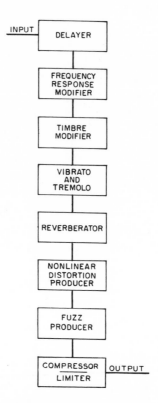

Fig. 15.5 An electronic modifier consisting of a delayer, frequency response modifier, timbre modifier, vibrato and tremolo modulator, reverberator, nonlinear distortion producer, fuzz producer, compressor and limiter.

is another example of the introduction of what is normally considered to be undesirable, namely, noise, to heighten the artistic aspects of the recorded sound.

The volume compressor of Fig. 15.5, which was described in Section 12.2, is an electronic system which reduces the amplification in a gradual manner when the signal input attains a certain level. Compressors are used to reduce the volume or amplitude range in sound motion picture, magnetic tape, and phonograph disk recording, and sound broadcasting, sound-reenforcing systems, etc. A reduction in the volume range in radio, magnetic tape, and phonograph sound reproduction makes it possible to reproduce the wide amplitude range of orchestra music in the home without excessive top levels. The use of the compressor improves the signal-to-noise ratio and thereby

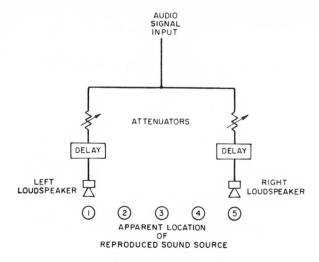

Fig. 15.6   A schematic diagram depicting delayers and attenuators for placing an apparent sound source of an audio signal at any point between two stereophonic loudspeakers reproducing the audio signal.

improves the intelligibility of speech and enhances the music sound reproduction when the ambient noise is high.

An extension of the tone modifier is shown in Fig. 15.7. The tone modifier of Fig. 15.7 makes it possible to extend the frequency range of an instrument and modify the timbre. The fundamental selector searches for and selects the fundamental frequency of the instrument. Then this fundamental frequency is multiplied or divided to extend the frequency range of the instrument or to produce more than one instrument, as depicted in Fig. 15.7. Under these conditions the original sound source or instrument supplies the fundamental frequency or pitch, the envelope, and the loudness.

The systems depicted in Figs. 15.5, 15.6, and 15.7 are very powerful tools for modifying the original tones and auditory perspective of the instruments as recorded in Figs. 15.2 and 15.3. The general idea is to provide new sounds that will increase the subjective response to the ultimate product, the record, when it is reproduced.

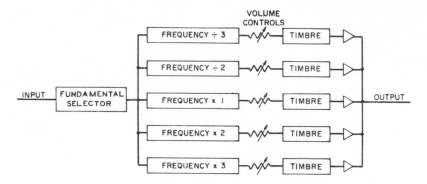

Fig. 15.7 A system for selecting the fundamental frequency of an audio signal and multiplying or dividing the frequency by multiples and injecting new timbre to produce several new instruments from the input signal of a single instrument.

The system of Fig. 15.4 converts eight channels of original sound to two-channel stereophonic sound. By the addition of two more sections of eight channels each as shown in Fig. 15.8, the eight channels may be allocated and recorded as four-channel quadraphonic sound. Reverberation envelope, sound in all manner of motion, and other spatial effects may be obtained in the reproduction in quadraphonic sound.

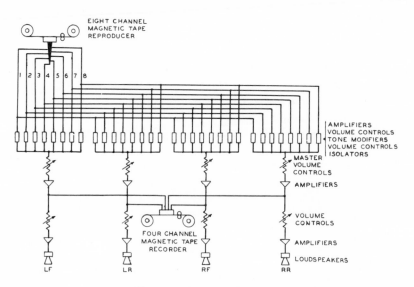

Fig. 15.8 The mixing, modifying, and conversion to four channel quadraphonic magnetic tape of the eight sound sources recorded as shown in Figs. 15.2 and 15.3.

15.4 SOUND MOTION PICTURE STUDIO (SOUND STAGE)

A sound motion picture studio, also termed a sound stage, is a large acoustic-ally treated enclosure used to house a setting and the equipment used in sound motion picture recording. The sound stage shown in Fig. 15.9 is equipped with air conditioning, catwalks, power outlets, and other facilities for the production of sound motion pictures. In the sound recording setup depicted in Fig. 15.9 the initial monitoring is carried out at a console located on the sound stage. The magnetic tape recorder synchronized with the camera is located at a remote point in a recording room, as is shown in the figure. (See Section 10.2D.) In the recording of sound motion pictures, the micro-phone must be kept out of the picture. This is done by suspending the micro-phone on a boom so that it can be raised or lowered and moved about by the boom engineer. (Fig. 15.9.) The boom is also equipped with a manually operated swivel arrangement so that the boom engineer can have the directional microphone pointed at the action. The cardioid-type unidirectional microphone is used for the recording of sound motion pictures. The sound monitoring is carried out with earphones. In some productions a small portable booth equipped with a large glass window is used to house the

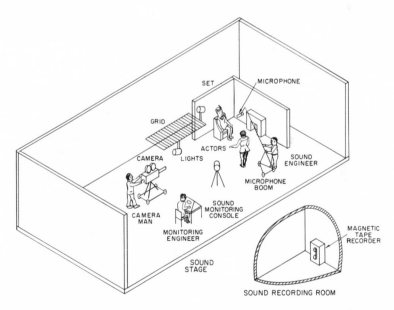

Fig. 15.9 Perspective view of a sound motion picture sound stage, equipment and personnel employed in filming and recording a sound motion picture.

monitoring equipment. The sound recording system shown in Fig. 15.9 provides a single channel. Many sound motion pictures are recorded in stereophonic sound, in which three channels are used for auditory perspective and the fourth for a surround channel. (See Section 16.6.) For stereophonic sound pickup of dramatic action, three spaced microphones are usually mounted on a single boom and connected to a three-channel recording system synchronized with the camera.

The music recording for sound motion pictures is usually carried out in a recording studio as described in the preceding section.

The sound pickup distances are sometimes very large in recording sound motion pictures. Therefore, the reverberation time of sound stages is made as low as possible. This expedient together with the use of directional microphones reduces the effective reverberation and permits long pickup distances.

Typical overall dimensions for a large studio are as follows: height, 45 feet; width, 100 feet; and length, 140 feet. A reverberation time of about one-half second is possible for stages with a volume of about 500,000 cubic feet. In the case of smaller stages a lower reverberation time may be obtained. It is usually standard practice to erect several sets on a single stage. This procedure may render some of the absorbing material of the stage ineffective and thereby increase the reverberation time. These undesirable effects may be overcome by the use of heavy sound-absorbing curtains that shield the different sets from each other.

15.5 TELEVISION STUDIO

A television studio is a large acoustically treated enclosure used to house the setting and equipment used in the picture and sound pickup for the telecasting of a television program. The enclosure is essentially of the same design as that used for sound motion pictures. The general arrangement of the performers, cameras, and microphones for a television broadcast is shown in Fig. 15.10. The microphone is mounted on a boom and kept out of the picture for the dramatic action. The microphones for the orchestra are usually not hidden. The monitoring room is located adjacent to the studio. A large window between the studio and the monitor room makes possible for the director, monitoring engineers, and recording engineer to view the action in the studio. Picture- and sound-monitoring equipment consists of an audio console, a video console, monitoring loudspeakers, monitoring kinescopes, and a magnetic tape recorder. Communication with the stage is carried out by means of the talk-back system shown in Fig. 15.10.

In many instances a television show is played to an audience. The general arrangement is the same as in Fig. 15.10, except that the audience is located in front, and the action is played to the audience.

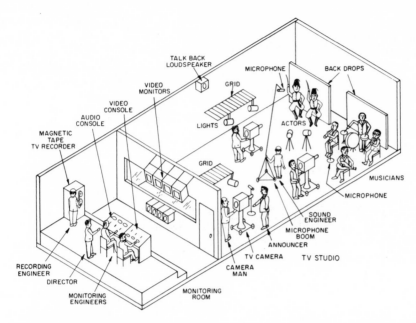

Fig. 15.10 Perspective view of a television studio, monitoring room equipment and personnel employed in telecasting.

Television shows are also recorded on magnetic tape for use in later telecasts and for future repeats.

The acoustic treatment, sound isolation, and general design of television studios are essentially the same as for sound motion studios described in the preceding section.

15.6 REVERBERATION TIME OF STUDIOS

As indicated in the preceding sections the reverberation time of studios for sound motion pictures and television is made very low. In general the set determines the acoustics of the action. For recording of music there is an optimum reverberation time. However, in recent times where the musicians are picked up separately, the tendency has been to use a value somewhat lower than the optimum value in order to provide the desired isolation. The reverberation time of recording studios as a function of studio volume is shown in Fig. 15.11. The reverberation time shown in Fig. 15.11 is lower than is usually recommended. However, since synthetic reverberation is usually introduced to augment the collected reverberation, the original reverberation is no longer important. (See Section 15.3.)

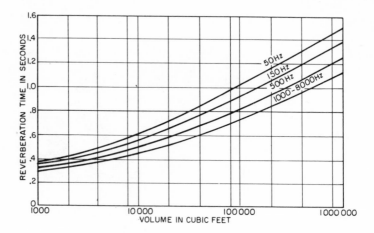

Fig. 15.11   Reverberation time characteristic of a recording or broadcasting studio as a function of the studio volume.

15.7 WALL AND CEILING ACOUSTIC TREATMENT FOR RECORDING STUDIOS

The main objective in recording studios is to produce a diffuse sound field free of standing wave systems in the reverberant sound. The problem of obtaining a uniform sound pressure distribution in the reflected sound can be solved by the use of wall and ceiling surfaces which diffuse, distribute, and disperse the sound reflected from the walls. Three typical wall and ceiling sound treatment designs for obtaining a diffuse and uniform sound distribution in the reflected sounds are shown in Fig. 15.12. In Fig. 15.12A the sound

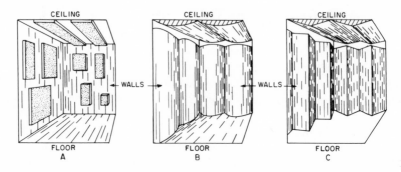

Fig. 15.12   Wall and ceiling structures for diffusing the reflected sound. A. Sound-absorbing material located in spots. B. Cylindrical surfaces. C. Serrated surfaces.

treatment is distributed in discrete spots on the walls and ceiling. This distribution of sound-absorbing material breaks up the wavefront of the reflected sound and thereby produces a diffuse sound field. In Fig. 15.12B the polycylindrical surface on the walls and ceiling increases the wavefront of the reflected sound and thereby produces a diffuse sound field. In Fig. 15.12C the serrated surface on the walls and ceiling introduces different path lengths for the reflected sound and thereby produces a diffuse sound field.

The use of the wall and ceiling treatment of Fig. 15.12 leads to a uniform sound decay curve and reduces or eliminates echos and flutters in the sound field. In some designs all three types of sound treatment are used in the same studio.

15.8 VARIABLE ACOUSTIC WALL TREATMENT

There are cases where the studio with a variable reverberation time is desired. This can be accomplished by a wall in which the acoustic absorption can be varied or by means whereby the volume is varied. A wall treatment in which the sound absorption can be varied is shown in Fig. 15.13. In Fig. 15.13A the highest reflection is obtained and thereby the longest reverberation time.

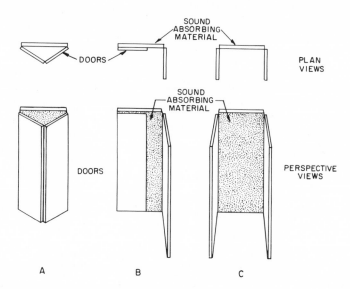

Fig. 15.13 Variable sound absorbing wall treatment. A. Highly sound-reflecting. B. Partially sound absorbing. C. Highly sound absorbing.

In Fig. 15.13B 30 percent of the highly sound-absorbing wall is exposed. In Fig. 15.13C the entire wall surface is highly absorbent. From condition 15.13A to condition 15.13C the sound absorption coefficient varies from 0.1 to 0.8. This wide variation in sound-absorption treatment provides the means for obtaining a reverberation time varying over a range of two-to-one.

15.9 VARIABLE STUDIO VOLUME

A wall treatment in which the acoustic absorption can be varied has been described in the preceding section. The reverberation time can also be varied by changing the volume of the studio. This is accomplished by a movable ceiling so that the height of the ceiling can be varied. The change in volume is usually of the order of two-to-one.

15.10 NOISE CRITERIA

The noise level in a studio should not be the limitation on the noise level of the reproduced sound. For the establishment of the permissible noise level in a studio the Noise Criteria (NC) are used. The Noise Criteria characteristics are shown in Fig. 15.14. For radio, recording, sound motion picture, and television studios the usual recommendation is that the NC value should not exceed 20. However, this is on the high side because with a noise level corresponding to an NC of 20, some of the ambient noise may be heard in the recording. Therefore, a noise level corresponding to an NC of 10 to 15 is considered to be more desirable. In addition, the noise should be uniform without spikes of noise that penetrate the noise criteria characteristics. (A desirable noise spectrum characteristic is shown later, in Fig. 17.12. The hearing limits for pure tones in a studio with a noise level corresponding to an NC of 10 is shown in Fig. 17.13.)

15.11 NOISE ISOLATION FOR STUDIOS

The desired ambient noise level in a studio was considered in Section 15.10. To achieve this order of ambient noise and vibration there must be considerable isolation from airborne sound and structural vibrations surrounding the outside of the studio.

The most troublesome high-level noise sources for which isolation must be provided are automotive and aircraft traffic. Measurements of the noise

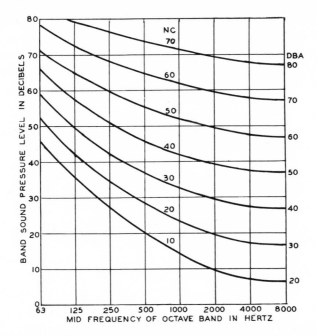

Fig. 15.14 The Noise Criteria (NC) characteristics for octave bands with the center frequencies as shown in graph and the frequency response of the octave filters as shown in Fig. 13.4. The approximate overall dBA noise level of the studio is depicted on the right of the graph. For example, an NC level of 20 in the octave bands as indicated leads to an overall noise level of 30 dBA.

levels at the location of the studio indicate the magnitude of the isolation required to achieve the low-order of ambient noise in the studio. (The exposition in Section 14.2D and Table 14.2 indicated the type of structure that would provide the required isolation.) Double-wall construction is almost imperative where the outside noise levels are high. The roof or ceiling presents the most difficult problem because here again very massive and double partitions must be used to isolate against aircraft noise. In the case of a multiple studio complex there must be isolation between the studios because the sound level in an operating studio may be very high indeed.

The floors of studios should be rigid and massive to prevent transmission of sound along the floor due to impacts, as for example, in the case of large dancing groups employed in sound motion picture and television studios. An improvement in the case of the floor can be effected by dividing the floor into sections and isolating each section mechanically. Mechanical filters and networks are used to reduce transmission of vibrations to the studio from outside-generated vibrations. This is a very complex subject and beyond the scope of this text.

15.12 NOISE IN A SOUND-COLLECTING SYSTEM

A sound-collecting system consisting of a studio, microphone and amplifier is depicted in Fig. 15.15. The noise levels depicted in Fig. 15.15 are all referred to the ambient noise in the studio. Furthermore, the overall gain is adjusted so that the ratio of the voltage output of the amplifier to the sound pressure input to the microphone is the same for both A and B of Fig. 15.15. The studio is assumed to exhibit a noise level corresponding to an NC of 10. This is a dBA level of 20. These noise levels are only experienced in exceedingly quiet studios. In general, the recommendations are that the noise in a studio should not exceed an NC of 20. The microphone noise level shown in Fig. 15.15A is for a very sensitive microphone employing electrodynamic transducer, as exemplified by the dynamic and ribbon types. The thermal noise generated in the microphone conductor can be derived from Equation (5.5). The microphone noise level shown in Fig. 15.15A is also representative of a very sensitive condenser microphone. The noise level in a properly designed amplifier should always be at least 5 dB below the noise level of the microphone. The thermal acoustic noise level of the air due to the thermal agitation of the molecules is well below all the other noises. In Fig. 15.15A the studio ambient noise level is 6 dB above the microphone noise level and therefore establishes the noise level in the system. The microphone noise level depicted

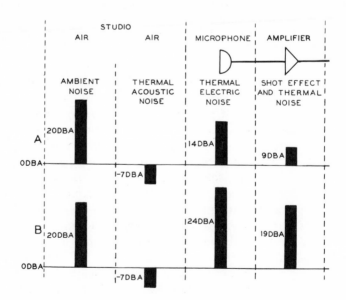

Fig. 15.15 The noise levels in a sound pickup system. A. Microphone with high sensitivity. B. Microphone of moderate sensitivity.

in Fig. 15.15B is for a microphone of moderate sensitivity, that is 10 dB above the level of the very sensitive microphone of Fig. 15.15A. Now the noise level in the microphone is 4 dB above the ambient noise level of the studio, and the microphone establishes the noise level in the system.

The data of Fig. 15.15 show that the microphone need not be the limitation on noise. As a matter of fact, a microphone has been developed and built in which the thermal noise in the air is the limitation. Sensitivity beyond that is of no significance from any standpoint.

REFERENCES

Olson, Harry F. *Acoustical Engineering*, Van Nostrand Reinhold, New York, 1957.

Olson, Harry F. *Music, Physics, and Engineering*, Dover Publications, New York, 1967.

Olson, Harry F. "Electronic Music Synthesis for Recordings," *IEEE Spectrum*, Vol. 8, No. 3, p. 18, 1971.

Olson, Harry F. "Microphone Thermal Agitation Noise," *Journal of the Audio Engineering Society*, Vol. 18, No. 3, p. 336, 1970.

Olson, Harry F. "Microphone Thermal Agitation Noise," *Journal of the Acoustical Society of America*, Vol. 51, No. 2, Part 1, p. 425, 1972.

Volkmann, John E. "Polycylindrical Diffusers," *Journal of the Acoustical Society of America*, Vol. 13, No. 3, p. 234, 1942.

Volkmann, John E. "Acoustic Requirements of Stereo Recording Studios," *Journal of the Audio Engineering Society*, Vol. 14, No. 4, p. 324, 1966.

Woram, John M. "A Pop Recording Session," *Audio*, Vol. 53, No. 5, p. 21, 1968, and Vol. 53, No. 6, p. 30, 1968.

(See also numerous articles on studio and recording techniques in the *Journal of the Audio Engineering Society*.)

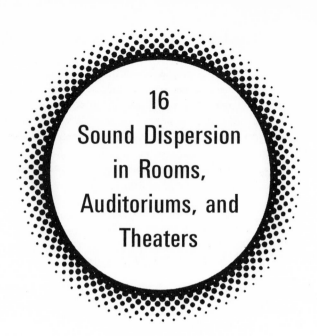

16
Sound Dispersion
in Rooms,
Auditoriums, and
Theaters

16.1 INTRODUCTION

The dispersion of sound in an enclosure was considered in Chapter 14. The fundamentals of a sound source operating in a room were developed in Section 14.2. The reproduction of sound takes place in residences, offices, schools, hospitals, factories, auditoriums, and theaters. The three main subdivisions are the reproduction of sound from some original sound source in the form of records and broadcasts, the reenforcement of a sound source in the same room, and the reproduction of an original sound source from an adjacent room. The purpose of this chapter is to provide an exposition of the reproduction of sound in enclosures under the above outlined conditions.

16.2 REVERBERATION TIME OF ROOMS, AUDITORIUMS, AND THEATERS

The reverberation time of an enclosure was developed in Section 14.2B. In order to provide a reasonable effective reverberation of the reproduced sound, the reverberation time of the room should be lower than that of a passive enclosure. If more reverberation is desired in the reproduced sound, as for example, in the case of music, synthetic reverberation may be added. The reverberation time of rooms, auditoriums, and theaters for the reproduction of sound as a function of the volume of the enclosure is shown in Fig. 16.1. In view of the frequency response of the ear as depicted later in Fig. 17.4, the general subjective opinion is that the reverberation time should increase with decrease of the frequency as is shown in Fig. 16.1. From a subjective standpoint the low frequency accentuation of reverberation leads to more uniform decay of the sound over the audio frequency range.

16.3 REPRODUCTION OF SOUND IN A RESIDENCE

Sound is reproduced in a residence by means of three different field-type systems, namely, monophonic, stereophonic, and quadraphonic. The fundamentals of the field-type sound-reproducing systems were considered in Chapter 6, in which the four fundamental conditions required to achieve realism in the reproduced sound were outlined in Section 6.1. The purpose of this section is to provide a description of monophonic, stereophonic, and quadraphonic sound-reproducing systems and the performance of these systems in a room in a residence.

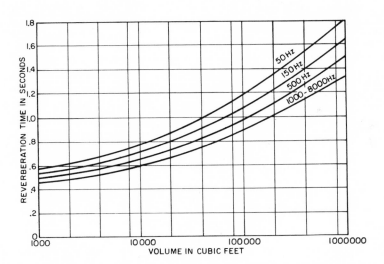

Fig. 16.1 Optimum reverberation time for rooms, auditoriums and theaters for the reproduction of sound.

A. Monophonic Sound-Reproducing System

A monophonic sound-reproducing system is of the field type in which an audio signal source, obtained from a disk reproducer, a magnetic tape reproducer, a radio receiver, or a television receiver, is reproduced by means of a loudspeaker as shown in Fig. 16.2A. The monophonic sound reproducing system will only satisfy conditions (1) and (2) on realism as described in Section 6.1.

Monophonic disk and magnetic tape reproducers are produced that range in size from very small portable models to the modules and integrated models described in the section which follows.

Monophonic radio receivers are of the amplitude and frequency modulation type ranging in size from the "small personal" to the table-model receivers. Practically no monophonic console radio receivers are produced today. The only examples are the modules and integrated models that will be described in the section which follows. Those instruments can reproduce monophonic amplitude and frequency modulation broadcasts with good fidelity. (See Sections 11.2 and 11.3.)

Television receivers provide monophonic sound reproduction. The large console television receivers are capable of reproducing sound with good quality from the standpoint of conditions (1) and (2) in Section 6.1.

B. Stereophonic Sound-Reproducing System

A stereophonic sound-reproducing system is of the field type in which an audio signal source, in the form of a disk or magnetic tape reproducer or an FM radio receiver, is reproduced by means of two loudspeakers as shown in

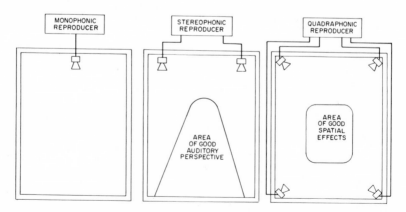

Fig. 16.2 Monophonic, stereophonic and quadraphonic sound reproducing systems.

Fig. 16.2B. The stereophonic sound-reproducing system will satisfy conditions (1), (2), and (4) on realism of Section 6.1.

Stereophonic sound reproduction provides auditory perspective of the reproduced sound and thereby preserves a subjective illusion of the spatial distribution of the original sound sources as described in Section 6.5 and depicted in Fig. 6.4.

There are two general classes of stereophonic sound reproducers, namely, module and integrated models.

The most versatile module system, from the standpoints of flexibility of disposition in the room and selection of the elements, is one in which each module represents a single element, as for example, an FM tuner, a record player or changer, a magnetic tape reproducer, an amplifier, and a loudspeaker. There are two general types of module reproducers: bookcase designs and console designs.

A bookcase type of module stereophonic sound reproducer, consisting of a tuner, a record player or changer, a magnetic tape recorder, an amplifier, and loudspeakers, is shown in Fig. 16.3. This module system is composed of relatively small and independent units that can be disposed in various ways in the room. Low frequency performance is determined to a large extent by the size of the loudspeaker cabinets.

The console type of module stereophonic sound reproducer shown in Fig. 16.4 provides some flexibility of disposition in the room combined with the finest performance. The loudspeakers, housed in relatively large cabinets, provide excellent low frequency response. The record player or changer, the magnetic tape recorder, the FM radio tuner, and the amplifier are housed in the center console.

The integrated stereophonic sound reproducer is one in which all the

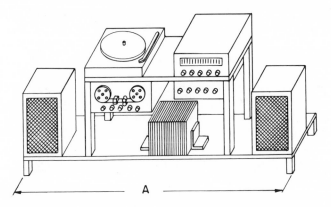

Fig. 16.3   Bookcase type of module stereophonic sound reproducing system.

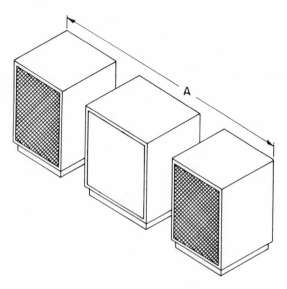

Fig. 16.4   Console type of module stereophonic sound reproducing system.

elements (record changer, magnetic tape reproducer, radio tuner, amplifier, and loudspeakers) are housed in a single large cabinet as shown in Fig. 16.5. The space allocated to the loudspeaker is relatively large and therefore provides excellent low frequency response.

The area over which stereophonic sound is reproduced with good auditory perspective is shown in Fig. 16.2B. The distance A in Figs. 16.3, 16.4, and 16.5 subtends about the same angle at the listening area of Fig. 16.2B as the expanse of the live source subtends at the listener in the range of seats in the concert hall. With good stereophonic sound pickup, the listener should be able to hear the various instruments and performers in the reproduced sound in the true original lateral locations of these sound sources.

C. Quadraphonic Sound-Reproducing System

A quadraphonic sound-reproducing system is of the field type in which an audio signal source, in the form of a magnetic tape reproducer, is reproduced on four loudspeakers as shown in Fig. 16.2C. The quadraphonic sound-reproducing system will satisfy all four conditions on realism as outlined in Section 6.1. In addition, as described in Section 6.1 the quadraphonic sound-reproducing system will reproduce sound in motion with all the manifold variations.

In general, the quadraphonic system depicted in the plan views of Figs.

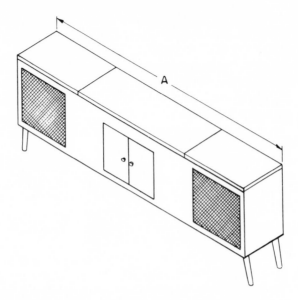

Fig. 16.5   Integrated type of stereophonic sound reproducing system.

6.5 and 6.6 is of the module type or console type as shown in Figs. 16.3, 16.4, and 16.5 with two additional channels and loudspeakers. The rear right and rear left loudspeakers are usually of the type shown in Fig. 16.3. However, the console-type loudspeaker shown in Fig. 16.4 may also be used and thereby provide better low frequency reproduction.

16.4 REPRODUCTION OF SOUND IN THE AUTOMOBILE

Radio receivers are installed in more than 75 percent of the automobiles sold today. Both amplitude and frequency modulation receivers are used in automobiles so that both monophonic and stereophonic sound reproduction may be obtained. Magnetic tape sound reproducers are also installed in automobiles. Monophonic, stereophonic, and quadraphonic sound reproduction may be obtained from the magnetic tape reproducer. The reproduction of sound in automobiles is a large factor in the mass dissemination of information and entertainment.

   In monophonic sound reproduction in the automobile, the loudspeaker is usually placed in a central location in the instrument panel or in the horizontal instrument cover behind the windshield as shown in Fig. 16.6. In some installations an auxiliary loudspeaker is located in the horizontal panel in front of the rear window as shown in Fig. 16.6.

In stereophonic sound reproduction in the automobile, the loudspeakers are usually located in the right and left top portions of the horizontal instrument cover behind the windshield as shown in Fig. 16.6. The stereophonic loudspeakers may also be located in the doors. In some installations two additional loudspeakers are located in the horizontal panel in front of the rear window as shown in Fig. 16.6.

In quadraphonic sound reproduction in automobiles, the front loudspeakers are usually located in the right and left top of the horizontal instrument cover behind the windshield and the rear loudspeakers are located in the horizontal panel in the front of the rear window.

Excellent auditory perspective is obtained in automobile sound reproduction from stereophonic and quadraphonic reproducers. The quadraphonic reproducer provides in addition to the auditory perspective, the reverberation envelope and spatial effects. The sound reproduction performance of the quadraphonic reproducer is outstanding in the automobile.

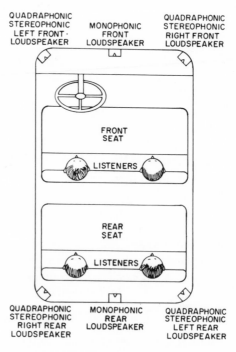

Fig. 16.6 The location of the loudspeakers in an automobile for monophonic, stereophonic and quadraphonic sound reproduction.

16.5 ACOUSTICALLY ACTIVE ARCHITECTURAL ENCLOSURE. SOUND-REENFORCEMENT SYSTEMS

"Acoustically passive architectural enclosure" is a general term used to designate auditoriums, theaters, concert halls, churches, classrooms, conference rooms, lecture halls, and other rooms employing passive acoustic means designed for the transmission of the sound of speech and music, produced by speakers or musicians, to the distributed listeners.

"Acoustically active architectural enclosure" is a general term used to designate auditoriums, theaters, concert halls, churches, classrooms, conference rooms, lecture halls, and other rooms employing active acoustic means designed for the transmission of the sounds of speech and music, produced by speakers or musicians, to the distributed listeners.

Prior to the advent of electronic sound reproduction the acoustically passive architectural enclosure was the only means available for the renditions of speech, voice, and music programs to large groups of listeners. As a consequence, considerable attention has been given to the science of architectural acoustics in the design of the acoustically passive architectural enclosure, with the object of achieving the maximum acoustic fidelity and the greatest artistic impact. Even the designs employing the highest quality of architectural acoustics leave much to be desired in performance. The many incompatible factors involved in the design of an acoustically passive architectural enclosure make it impossible to satisfy all the criteria for a high order of excellence of performance of the varied programs, renditions, and aggregations for all locations in the listening areas. Therefore, except in a very few instances, the acoustically passive architectural enclosure is rapidly disappearing from the modern scene. The main reason for the passing of the acoustically passive architectural enclosure is its limited, antiquated, inadequate, capricious, and atrocious acoustical performance as compared to the acoustically active architectural enclosure. Furthermore, new electronic developments have made the acoustic possibilities and properties of the acoustically active architectural enclosure practically universal and unlimited.

The electro-acoustic means employed in the acoustically active architectural enclosure are termed sound-reenforcement systems. The purpose of this section is to describe typical examples of the use of sound-reenforcement systems.

Two simple sound-reenforcement systems are shown in Fig. 16.7. The sound-reenforcement system consists of a microphone, amplifier, and loudspeaker as described in Section 7.4. In Fig. 16.7A the loudspeaker is located above the stage. The loudspeaker may be of the horn type described in Section 2.3 or the column type described in Section 2.2J. In Fig. 16.7B the loudspeakers are distributed around on the walls of the room or on the

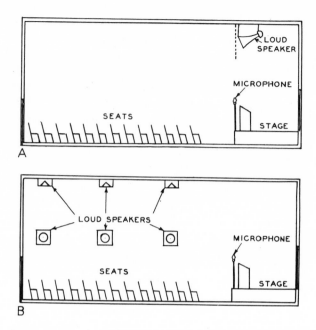

Fig. 16.7   Arrangements of the loudspeakers for a sound-reenforcing system in a large room as, for example, a schoolroom, a church, an auditorium, etc. A. A horn loudspeaker cluster located above the stage. B. A large number of direct radiator dynamic oudspeakers distributed on the walls and ceiling of the room.

ceiling or both. The systems of Fig. 16.7A or B may be employed with equally good results. The single loudspeaker station located above the original sound source produces somewhat more realistic results since the original and augmented sound originate in the same general direction. In the system with distributed loudspeakers it is somewhat easier to obtain good reenforcement because feedback difficulties are reduced. The latter arrangement is used for sound reenforcement in schoolrooms and churches.

   In the preceding examples the design of the auditorium is not coordinated with the sound-reenforcing system. The sound-reenforcing system is usually installed after the auditorium is built. The overall performance of the auditorium and sound system is not as good as it would have been if the design of the elements of the entire acoustical project had been coordinated. Obviously, the logical procedure is to design the auditorium and sound-reenforcing apparatus as an integrated system.

   A design of an acousto-electronic auditorium is shown in Fig. 16.8. The microphones used to pick up the sound on the stage are hidden behind a perforated ceiling. The loudspeakers are also hidden behind a perforated

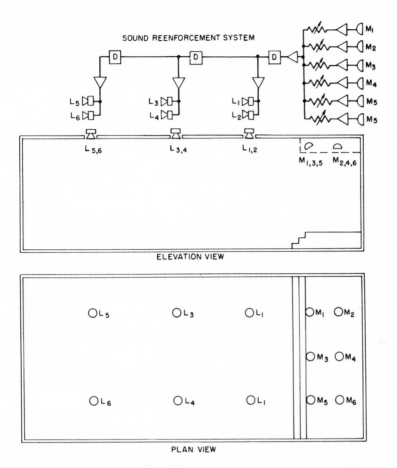

Fig. 16.8 Elevation and plan view of an electroacoustic auditorium with a sound reinforcement system with progressive time delay between the loudspeakers and the microphones. The block diagram depicts the elements of the sound reinforcement system.

grill in the ceiling. In order to provide a low effective reverberation in the reproduced sound, the reverberation time of the auditorium is on the low side of the reverberation time specified for auditoriums. (See Section 16.2.) To establish the illusion that all sound originates on the stage, progressive delay (described in Section 12.7) is introduced as shown in Fig. 16.8. The delay is selected so that the direct sound that emanates from the stage arrives at the listener before the reenforced sound. The loudspeakers are located in the ceiling and arranged so that complete sound coverage of the audience is obtained. Six microphones are used to pick up the sound on the stage. Second-order gradient microphones are used in order to provide a high ratio of direct to reflected sound pickup. (See Sections 4.6A and 14.3B.) The performer can walk all around the stage without being encumbered by a microphone with a cord or a wireless microphone.

The sound reenforcement system is operated so that the sound level in the seating area is about equal to the direct sound from the person speaking at a distance of 3 feet. The speaker can be picked up over the entire area of the stage with no appreciable variation in level or quality. The convenience and mobility afforded by not having to wear a personal microphone are out-standing and desirable features. Continuous manual monitoring is not required—the controls can be set and the sound-reenforcement system will operate without any attention. The combination of a high-quality, low-level sound-reenforcement system, a constant efficiency of sound pickup over the stage area, a uniform distribution of reproduced sound in the auditorium and a progressive delay system provides a sound-reproducing system in which one is not aware that the original sound is reenforced.

A highly sophisticated universal acousto-electronic auditorium that will provide the appropriate acoustics for all manner of programs ranging from speech to musical programs of all kinds is shown in Fig. 16.9. The seating capacity of the architectural enclosure of Fig. 16.9 is 1800 persons. The cubical volume is 380,000 cubic feet. The reverberation time of the empty auditorium in the mid-frequency range is less than one second. The reverberation time of the stage in the mid-frequency range is about one-half second.

The large number of loudspeakers in the audience area provide a very uniform sound distribution. The sound on the stage is picked up by directional microphones. The five sets of microphones are connected through amplifiers and delayers to the five loudspeakers, located in the proscenium arch, to supply the auditory perspective. The delay system is designed so that the sound appears to originate from the source on the stage. (See Section 12.7.) Electronic synthetic reverberators supply the reverberation envelope for the sound reproduced from the stage (see Section 12.6). The electronic synthetic reverberators provide reverberation times from 0 to 5 seconds. For speech the system is operated without any synthetic reverberation. Under these conditions the overall ratio of the direct to reflected sounds for the sound picked up on the stage and the sound reproduced in the audience area is 3 to 1. This corresponds to a very low reverberation time of $\frac{1}{12}$ second. The level of the speech at the listeners' ears is maintained at 78 dB, which corresponds to the level of conversational speech at 3 feet. The high ratio of direct to reflected sound combined with the normal level of the reproduced speech leads to practically perfect intelligibility and naturalness of speech. For every type of musical rendition there is an optimum reverberation time, as depicted in Table 16.1. The reverberation time of the ordinary passive

TABLE 16.1

Effective Reverberation

Type of Music	Time in Seconds
Organ	2 to 5
Band	2 to 3
Symphony	1.5 to 2.5
Chorus	1 to 2
Opera	0.7 to 1
Popular orchestra	0.5 to 1

architectural enclosure can be correct for only one type of musical rendition. The acousto-electronic system of Fig. 16.9 provides the correct reverberation time for all types of musical renditions. The complete flexibility of the system makes it possible to vary the reverberation during a rendition. Furthermore, the acoustics are the same for each and every seat location in the entire architectural enclosure.

16.6 SOUND MOTION PICTURE REPRODUCTION IN THE THEATER

An installation of a sound motion picture sound-reproducing system in a sound motion picture theater is shown in Fig. 16.10. A sound motion picture reproducing system was described in Sections 10.4A and 10.5C. The elements of the motion picture film sound reproducer constitute the sound head of the motion picture projector. (See Sections 10.4C and 10.5C.) The motion picture projector, amplifier, volume controls, and monitoring loudspeaker are located in the projection booth. The theater loudspeakers are located behind the screen.

The reverberation time of the theater should correspond to that given by Fig. 16.1. Care should be taken to avoid echoes and focusing of the reflected sounds by the use of convex wall and ceiling sections as shown in Fig. 16.10. (See Section 15.7.)

In a theater free of acoustic defects, the energy density of the generally reflected sound is practically the same for all parts of the theater. Therefore, the solution to the problem of achieving uniform sound energy density is to employ directional loudspeakers arranged high above the stage so that the direct sound energy density is the same for all parts of the theater.

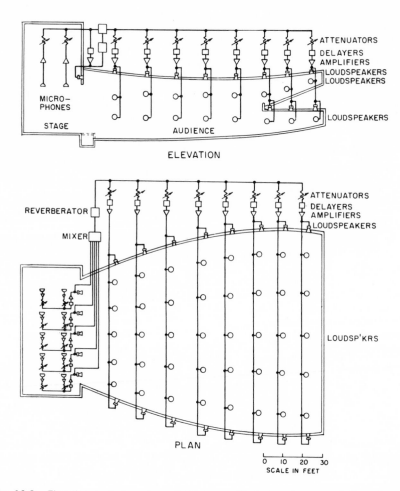

Fig. 16.9 Elevation and plan view of a large auditorium and concert hall equipped with a sophisticated acousto-electronic system consisting of microphones, delayers, reverberators, volume controls, amplifiers and loudspeakers.

For monophonic sound reproduction a single theater loudspeaker is located behind the screen, as shown in Fig. 16.10. Stereophonic sound is reproduced on three channels; the three theater loudspeakers representing the three channels are located behind the screen and reproduce the sound in auditory perspective as shown in Fig. 16.10. The loudspeakers located around the walls of the theater operate from a fourth channel and are used to reproduce the reverberation envelope and various sound effects that enhance the artistic aspects of the reproduced sound.

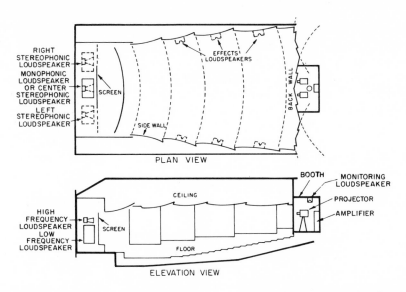

Fig. 16.10 Elevation and plan view of a sound motion picture theater showing the elements of the sound reproducing system. For monophonic sound reproduction, the center loudspeaker is used. For three-channel stereophonic sound reproduction, three loudspeakers are used. The loudspeakers located around the theater are used for sound effects and the reverberation envelope.

16.7 GENERAL ANNOUNCE AND PAGING SOUND-REPRODUCING SYSTEMS

General announce systems are useful in factories, warehouses, railroad stations, airport terminals, and so forth. A typical installation is depicted in Fig. 16.11A. For this type of work intelligibility is more important than quality. The deleterious effect of reverberation upon articulation can be reduced, and a better control of sound distribution can be obtained, by reducing the low frequency response of the system. Furthermore, the cost of the amplifiers and loudspeakers is also reduced by limiting the frequency range. To find the power required, the sound intensity level under actual operating conditions should be determined. The system should be designed to produce an intensity level 20 to 40 dB above the general noise level. Under no conditions should the system be designed to deliver an intensity level of less than 80 dB at the listener. The loudspeakers should be selected and arranged following an analysis similar to that outlined in the preceding sections, so that uniform sound distribution and adequate intensity levels are obtained.

For certain types of general announce, paging, and sound distributing installations used in schools, hospitals, department stores, hotels, and so on, the intensity level required is relatively low and the volume of the average room is usually small. For most installations of this type, save in noisy locations, an intensity level of 70 dB is more than adequate. Higher intensity levels tend to produce annoyance in adjacent rooms. Since the sound levels are low, the power requirements for the loudspeakers will be small. So that they blend with the furnishings of the room, it is desirable to mount the loudspeakers flush with the wall surface. Therefore, for these applications, a direct radiator loudspeaker of the permanent magnet dynamic type is most suitable. In this connection it should be mentioned that these loudspeakers have a very low efficiency, of the order of 1 percent, as compared to 25 percent to 50 percent for the horn loudspeakers.

For large rooms requiring large acoustical outputs it is more economical to use a high-efficiency loudspeaker and effect a corresponding reduction in the power amplifier requirements. On the other hand, for an installation of the type depicted in Fig. 16.11B and requiring a large number of units, it is more economical to use relatively inefficient low-cost loudspeakers and correspondingly larger amplifiers.

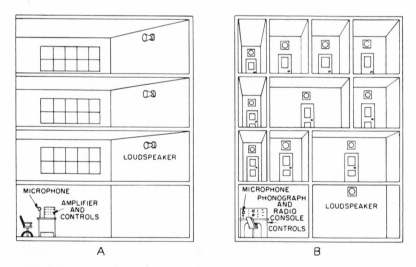

Fig. 16.11 Two examples of call, general announce and sound distributing systems. A. A high efficiency horn loudspeaker is used to obtain a high sound level over a large floor area as in a factory or warehouse. B. Small direct radiator loudspeakers are used to supply the small rooms at a relatively low level, as in paging, announcing, and centralized sound installations used in hospitals, hotels and schools.

16.8 POWER REQUIREMENTS FOR SOUND-REPRODUCING SYSTEMS

The power requirement and sound power level are important factors in home-type, motion picture theater, and sound-reenforcement systems. The minimum average or operating sound level which some of these systems should be capable of delivering is 80 dB at the listener (0 dB = 0.0002 microbar). The graph of Fig. 16.12 shows the acoustical power required to produce sound levels of 70, 80, 90, and 100 dB. In large auditoriums where the orchestra or organ is reenforced or reproduced, the maximum sound level should be at least 100 dB in order to render full artistic appeal. The sound level requirements for the various systems under the different conditions of operation have been discussed in the preceding sections and will not be repeated here. The amplifier output power needed will be at least ten times the acoustic power depending upon the efficiencies of the loudspeaker.

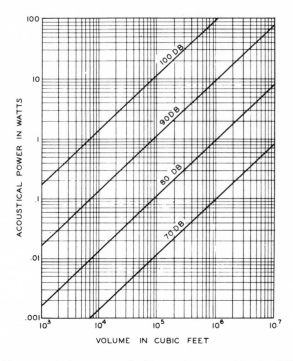

Fig. 16.12 The acoustical power required to produce sound levels of 70, 80, 90, and 100 decibels in a typical room, auditorium or theater as a function of the volume of the enclosure.

16.9 OUTDOOR SOUND-REENFORCEMENT SYSTEM

Regardless of the size of the athletic field or baseball park, a sound-reenforcement system, usually termed a public address system, is useful for informing those in the stands of what is happening on the field. In general, the chief requirements are as follows: uniform distribution of sound intensity in all parts of the stand, adequate power to override any anticipated noise level of the maximum crowd, and facilities available for microphones at predetermined points.

A large stadium equipped with a public address system is illustrated by Fig. 16.13A. Because of the size and configuration of the audience area it is practically impossible to obtain satisfactory sound level and coverage with a single loudspeaker. Consequently, the loudspeakers are placed at intervals near the boundary of the field sufficiently close together so that uniform response is obtained in the horizontal plane. The elevation view of Fig. 16.13A shows how uniform sound distribution is obtained in the vertical plane by means of the directional characteristics.

A baseball field equipped with a public address system is illustrated by Fig. 16.13B. As contrasted to the stadium, here a single loudspeaker station is used to supply the entire audience area. The distance between the loud-speakers and the auditors is very large. Therefore, the vertical coverage angle is very small, which means that practically any system will have a distribution

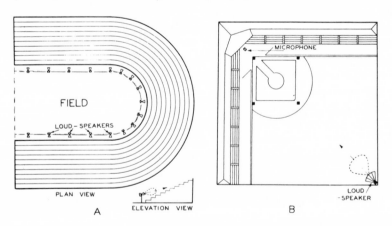

Fig. 16.13 Two arrangements of sound systems for addressing assemblages in large grandstands. A. For the stadium a large number of loudspeakers are used each covering a small portion of the total area. B. For the ball park a single loudspeaker cluster is used to supply the entire grandstand.

angle sufficiently broad to supply the required vertical spread. However, for conservation of power the vertical spread of the loudspeaker should correspond to the vertical angle subtended by the audience at the loudspeaker. Since the distance of those nearest the loudspeaker to those farthest removed (that is, considering the vertical angle only) is very nearly the same, the sound intensity from the loudspeaker will be practically the same for all parts of the audience through any vertical plane, and the use of compensation by means of the directional characteristics for change in distance in the vertical plane to obtain uniform response is not necessary. In the horizontal plane the spread of the loudspeaker should correspond to the angle subtended by the stands at the loudspeaker. Since the center line is farthest removed, a directional characteristic of the shape shown is necessary for obtaining the same sound level in all parts of the grandstand. To eliminate any difficulties due to feedback, directional microphones are used and oriented to reduce the effects of feedback.

The sound level required for public address work of the type considered above will be determined by the noise level of the maximum crowd. In general, it is not practical to employ a system with sufficient power to override the sound level during cheering, applause, and so on. However, the power should be sufficient to override the general noise during relatively quiet intervals. The noise level may be determined by means of a sound level meter. The power available should be sufficient to produce a minimum sound level of 80 dBA or, for very noisy conditions, 20 to 30 dB above the noise during the relatively quiet intervals. In the two examples cited above, and, in fact, for all outdoor public address work, the only consideration is direct sound. The problem is to select amplifiers and loudspeakers with characteristics that will deliver the required sound level over the distances and areas considered. The steps in the selection of a system may be as follows: First, the directional characteristics should be determined, as outlined in the preceding discussion, so that uniform response is obtained over the audience area. Second, either a single loudspeaker or a group of loudspeakers having the desired directional characteristics should be selected. Third, the response characteristic of the system on the axis at a specified input and distance should be available to show the amplifier power required to supply the desired intensity level. Fourth, the power-handling capacity of the loudspeakers and amplifiers should be adequate to supply the required intensity level.

Two types of sound-reenforcement installations for an outdoor theater are shown in Fig. 16.14. The system employing a single loudspeaker station located either above or below the stage is shown in Fig. 16.14A. If the stage is quite low, the logical position for the loudspeakers is at the top of the stage. The procedure for obtaining uniform sound coverage and adequate intensity

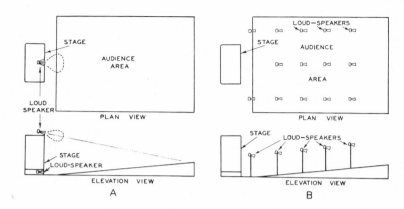

Fig. 16.14 Two arrangements of the sound reenforcement systems for an outdoor theater. A. A single loudspeaker is above the stage and has suitable directional characteristics to produce a uniform sound level over the audience area. For an alternate arrangement the loudspeaker is located below the stage floor. B. A number of loudspeakers are used each covering a small portion of the audience.

level of the sound from the loudspeakers used in the preceding considerations is applicable in this case. If the stage is very high, the separation between the loudspeakers and the action on the stage will be very large. As a consequence, the wide difference in the direction of the direct sound and the reenforced sound will be particularly disconcerting to listeners in the front portion of the seating area. Under these conditions, it may be desirable to locate an additional loudspeaker under the stage, as shown in Fig. 16.14A. The system depicted in Fig. 16.14B employs a large number of loudspeakers, each one supplying a small portion of the audience. The directional characteristics of the loudspeakers should be selected so that each individual area is adequately supplied. Cognizance must be taken of the energy supplied from adjacent loudspeakers.

There are certain advantages in each system. In the case of the single loudspeaker system, better illusion is obtained because the augmented sound appears to come from the stage. On the other hand, the intensity level outside the audience area in a backward direction falls off very slowly. At a distance equal to the length of the audience area the level is only 6 dB lower than that existing in the audience area. In certain locations the sound levels produced by such systems will cause considerable annoyance to those located in the vicinity of the theater. If the theater area is divided into small plots, each supplied by a loudspeaker, and the loudspeakers are directed downward, the sound intensity level outside the audience area will be considerably lower than it is in the case of the single loudspeaker station and usually will eliminate any annoyance difficulties. The short sound-projection distance is another advantage of the multiple loudspeaker system.

The above typical examples of outdoor public address and sound, reenforcement systems illustrate the principal factors involved in that field of sound reproduction.

16.10 OUTDOOR SOUND MOTION PICTURE THEATER

A "drive-in" motion-picture theater is an outdoor system in which the audience attends the show while seated in automobiles. A type of sound motion-picture drive-in theater system is shown in Fig. 16.15. Sound motion-picture reproducing systems are described in Sections 10.4 and 10.5C. The phototube and optical system are housed in the picture projector of Fig. 16.15. The picture projector, voltage power amplifiers, and master volume controls are located in the projection booth. In the drive-in theater, the customers are seated in their automobiles to view the picture and hear the

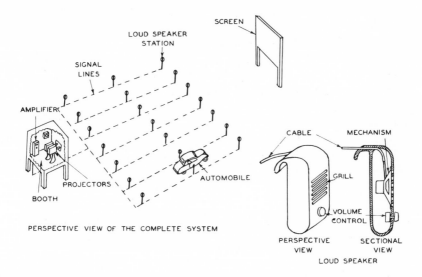

Fig. 16.15 A perspective view showing the elements of a drive-in sound motion picture threater. Perspective and sectional views of the loudspeaker used in the drive-in sound motion picture theater are shown in the lower right.

sound. Individual loudspeakers are provided to supply the sound to each automobile. Perspective and sectional views of such a loudspeaker are shown in Fig. 16.15. In use, the loudspeaker is hung inside the automobile. A volume control incorporated with the loudspeaker makes it possible for the listener to adjust the level of the reproduced sound. When not in use the loudspeakers are hung on posts provided for the purpose. Signal lines connect the individual loudspeakers to the amplifier in the booth.

16.11 NOISE LEVELS

The ambient noise level in residences, offices, schools, hospitals, factories, auditoriums, and theaters influences the intelligibility of low-level reproduced speech and places a lower limit on the sounds that can be heard in reproduced music. The ambient noise level in a well-designed studio should be from 10 to 30 dBA. The ambient noise level in a well-designed auditorium and theater in an empty condition should be from 25 to 35 dBA. With an audience the level will be around 47 dBA. The ambient noise level in the average residence is 30 to 42 dBA. The ambient noise level in a typical business office in a large city is 45 to 55 dBA. In the average factory the ambient noise level is 50 to 75 dBA. The spectrum levels of the ambient noise in the factories, offices, theaters, residences, and studios are shown in Fig. 17.12. The hearing limits for pure tones for these locations and the ambient noise levels indicated above, are shown in Fig. 17.13.

Among the major ambient noise offenders are heating and air conditioning systems. Great care and considerable skill are required in the acoustic design of these systems to obtain a low noise level.

The noise levels (dBA) and the noise criteria (NC) for various locations and conditions are shown in Table 16.2. The values shown in Table 16.2 are typical but there may be wide variations from the values shown in the table. The subject of noise criteria (NC) was discussed in Section 15.10, and noise criteria contours are depicted in Fig. 15.14.

16.12 PREFERRED LEVEL OF SOUND REPRODUCTION AND ROOM VOLUME

There is a very difinite relationship between the preferred or tolerable top level of sound reproduction and the volume of the room. Subjective tests have been carried out on the preferred or tolerable top level of sound reproduction

TABLE 16.2

Noise Levels for Various Sources and Locations

Source or Description of Noise		Noise Level dBA	Noise Criteria NC
Threshold of pain		130	120
Hammer blows on a steel plate 2 ft.		114	104
Riveter 35 ft.		97	87
Factories and shops		50–75	40–65
Busy street traffic		68	58
Ordinary conversation 3 ft.		65	55
Railroad station		55–65	45–55
Airport terminal		55–65	45–55
Stadiums		55	45
Large office		60–65	50–55
Factory office		60–63	50–63
Large stores		50–60	40–50
Medium store		45–60	35–50
Restaurant and dining rooms		45–55	35–45
Medium office		45–55	35–45
Automobile 50 m.p.h.		45–50	
Garage		55	45
Small store		45–55	35–45
Hotel		42	32
Apartment		42	32
Home in large city		40	30
Home in the country		30	20
Motion picture theater empty	For	25–35	15–25
Auditorium empty	Full	25–35	15–25
Concert hall empty	Add	25–35	15–25
Church empty	5 to 15	30	20
Classroom empty	dB	30	20
Broadcast studio. No audience		20–25	10–15
Television studio. No audience		25–35	15–25
Television studio. Audience		30–40	20–30
Sound motion picture stage		20–35	10–35
Recording studio		20–30	10–20
Average whisper		15–20	5–10
Quiet whisper 3 ft.		10–15	0–5
Threshold of hearing		0–5	0

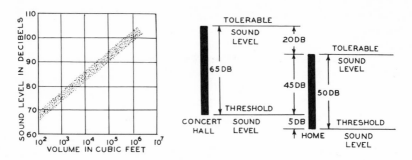

Fig. 16.16 The relation between the tolerable top-level of sound reproduction and the room volume. The bar graphs depict the tolerable top levels and average threshold levels in the concert hall and the home as well as the amplitude ranges in the two locations. The average threshold level for hearing pure tones in the home is 30 dB and in the concert hall 35 dB. (See Section 17.12.)

and the volume of the room. The results are shown in Fig. 16.16. The top level in the average room in the home is 80 dB. For the concert hall the top level is 100 dB. Referring to Fig. 17.13 the threshold due to the ambient noise is about 30 dB for the home and the threshold due to the ambient noise in the concert hall is about 35 dB. The bar graphs depict the threshold levels and tolerable top levels for the home. The bar graphs also show that the amplitude ranges in the home and the concert hall are 50 dB and 65 dB respectively.

REFERENCES

Beranek, Leo L. *Music, Acoustics, and Architecture*, Wiley, New York, 1962.

Boner, C. P. and Boner, C. R. "Behavior of a Sound System Response Immediately Below Feedback," *Journal of the Audio Engineering Society*, Vol. 14, No. 3, p. 200, 1966.

Brinkerhoff, D. E. and Schwarz, B. A. "Some Observations on Reproduced Sound in an Automobile," *Journal of the Audio Engineering Society*, Vol. 6, No. 1, p. 58, 1958.

Olson, Harry F. *Acoustical Engineering*, Van Nostrand Reinhold, New York, 1957.

Olson, Harry F. "Acoustoelectronic Auditorium," *Journal of the Acoustical Society of America*, Vol. 31, No. 7, p. 872, 1959.

Olson, Harry F. "Passive and Active Acoustics in Architectural Enclosures," *Journal of the Audio Engineering Society*, Vol. 13, No. 4, p. 307, 1965.

Rettinger, Michael. *Acuostics, Room Design, and Noise Control*, Chemical Publishing Co., New York, 1968.

(See also numerous articles on the acoustics of rooms, theaters, auditoriums, and sound-reenforcement systems in the *Journal of the Audio Engineering Society* and the *Journal of the Acoustical Society of America*.)

17
Subjective
Acoustics

17.1 INTRODUCTION

The major portion of this book has been concerned with expositions on the theory, design, construction, and testing of acoustical apparatus for the reproduction of sound. The ultimate significant destination of all reproduced sound is the human ear. Therefore, the subjective aspects of reproduced sound are important factors in any sound-reproducing system. Reproduced sounds may be classed as speech, music, and noise. The response of the human hearing mechanism to these reproduced sounds constitutes an important element in the ultimate performance of sound-reproducing equipment. The purpose of this chapter is to provide a description of the salient subjective factors of hearing involved in, and related to, the reproduction of sound.

17.2 HUMAN HEARING MECHANISM

The human hearing mechanism, shown in Fig. 17.1, may be divided into three parts: the outer ear, the middle ear, and the inner ear. The outer ear consists of the external ear or pinna and the ear canal, which is terminated in the eardrum or tympanic membrane. Behind the eardrum is the middle ear, a small

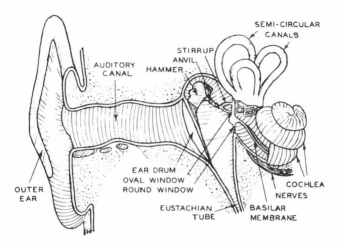

Fig. 17.1   Sectional and perspective views of the human hearing mechanism.

cavity in which three small bones—the hammer, the anvil, and the stirrup—form the elements of a system for transmitting vibrations from the eardrum to an aperture, termed the oval window of the inner ear. The cavity in the middle ear is filled with air by means of a pressure-equalizing tube, termed the Eustachian tube, leading to the nasal pharynx. The casing of the inner ear or cochlea is a bony structure of a spiral form (two-and-three-quarter turns). The cochlea is divided into three parts by the basilar membrane and Reissner's membrane. These three parallel canals are wound into the spiral. On one side of the basilar membrane is the organ of Corti, which contains the nerve terminals in the form of small hairs extending into the canal of the cochlea. These nerve endings are stimulated by the vibrations in the cochlea.

A schematic cross-sectional view of the ear and the acoustical network of the vibrating system are shown in Fig. 17.2. When a sound wave impinges upon the ear, it enters the ear canal and causes the eardrum to vibrate. The vibration of the eardrum is transmitted to the inner ear or cochlea by the three bones of the middle ear. The cochlea may be considered to be made up of distributed constants as shown in Fig. 17.2. The volume currents in the branches are indicated as $U_1, U_2, \ldots, U_K$ in branches $1, 2, \ldots, K$. These volume currents in turn actuate the nerves. High frequency sounds excite the portion of the cochlea nearest the oval window as shown in Fig. 17.2. Low frequencies are associated with the extreme end removed from the oval window. In other words, the cochlea is a frequency-discriminating system in which a certain vibration frequency is associated with a certain definite section of the cochlea. The auditory nerves that terminate all along the cochlea are stimulated by the vibrations. The activated nerve section sends

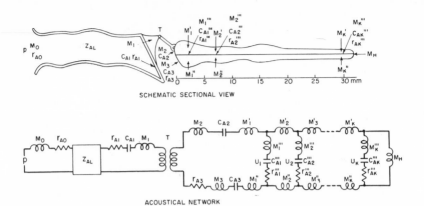

SCHEMATIC SECTIONAL VIEW

ACOUSTICAL NETWORK

Fig. 17.2 Schematic sectional view and acoustical network of the human hearing mechanism. In the acoustical network: M_0 and r_{AO} = inertance and acoustical resistance of the air load upon the opening to the ear canal. z_{AL} = four terminal acoustical network representing the ear canal. M_1 = inertance of the ear drum and hammer. C_{A1} and r_{A1} = acoustical capacitance and acoustical resistance of the ear drum. T = transformer consisting of the hammer, ossicles, anvil, arm and stirrup. M_2 = inertance of the stirrup and oval window. C_{A2} = acoustical capacitance of the oval window and tensor stapedius. M_3, C_{A3} and r_{A3} = inertance, acoustical capacitance and acoustical resistance of the round window. $M'_1, M'_2, \ldots, M'_K$ = inertances of the liquid in the scala vestibula $M''_1, M''_2, \ldots, M''_K$ = inertances of the liquid in the scala tympani $M'''_1, M'''_2, \ldots, M'''_K$, $r'''_{A1}, r'''_{A2}, \ldots, r'''_{AK}$, and $C'''_{A1}, C'''_{A2}, \ldots, C'''_{AK}$ = inertances, acoustical resistances and acoustical capacitances of the basilar membrane which separates the upper from the lower liquid. M_H = inertance of the liquid in the helicotrema. The volume currents $U_1, U_2, \ldots,$ U_K = volume currents in the nerve terminals.

a pulse to the brain which in turn is translated into a definite pitch. The frequency depends upon the nerves which are actuated. The greater the intensity of the sound the greater is the excitation of the hair cells, and a correspondingly greater number of nerve impulses are sent to the brain. The loudness as perceived by the brain is proportional to the number of nerve impulses. The nerve fibers that connect the cochlea to the brain are depicted in Fig. 17.3. In the ear mechanism as shown in Figs. 17.2 and 17.3, the cochlea is in effect a sound analyzer. The various frequencies in a complex sound which impinge upon the ear are sorted out by the frequency-selective properties of the cochlea. The exact nature of the frequency-selecting system is not too well known or understood. Unlike the relatively slow response of electrical and electronic systems with a high order of frequency selectivity, the response of the ear is so fast that only a few cycles of a sound wave are required to bring the hearing mechanism to full sensitivity. The ear is capable of distinguishing 1500 separate frequencies. No known electronic system with the rapid response of the ear is capable of resolving this number of discrete

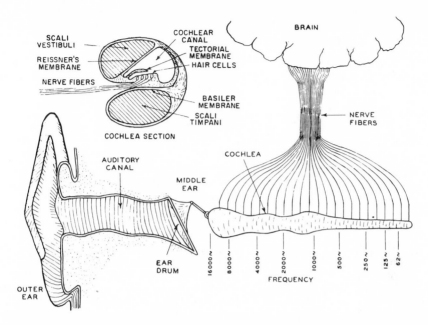

Fig. 17.3 Schematic view of the human hearing mechanism showing the outer ear, the middle ear, the cochlea, the nerve fibers leading to the brain and a section of the cochlea. The frequencies associated with each part of the cochlea are shown in the schematic view of the cochlea.

frequencies in the audio frequency range. Since the length of the frequency-selecting system of the cochlea is 1.4 inches, each section of frequency response can be considered to be confined to a length of less than one-thousandth of an inch.

17.3 FREQUENCY RESPONSE CHARACTERISTICS OF THE HUMAN HEARING MECHANISM

The loudness of a pure tone depends upon the frequency and intensity. The relation is revealed in the equal-loudness contours of Fig. 17.4. The 1000-hertz tone is the reference frequency in these determinations. The frequency response characteristics shown in Fig. 17.4 are obtained with the sound source directly in front of the listener. The sound reaches the listener in the form of a plane progressive sound wave. The sound pressure level is measured in the absence of the listener. The listening is binaural, and the listeners are otologically normal persons in the age group 18-to-25 years. The dashed charac-

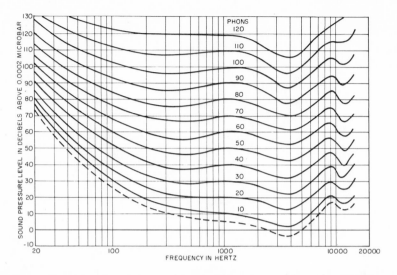

Fig. 17.4 Contour lines of equal loudness for binaural free-field listening conditions. Numbers on curves indicate loudness level in phons. Dashed line is the normal binaural minimum audible field (MAF). 0 dB = 10^{-16} watts per square centimeter. 0 dB = 0.0002 microbar. I.S.O. Recommendation R.226.

teristic depicts the normal binaural minimum audible sound, that is, the threshold of binaural hearing.

The frequency response characteristics of Fig. 17.4 show that if sound is reproduced at a level lower than the original sound level, some accentuation of the low frequency response is required to obtain the proper amplitude-frequency balance of the reproduced sound. To compensate for the difference in amplitude-frequency balance that is due to the lower reproduction sound level, an acoustically compensated volume control is used to accentuate low frequency response as the sound level of reproduction is lowered.

The frequency response characteristics of Fig. 17.4 are considered in the measurement of noise. The frequency response characteristics of the sound level meter are adjusted to correspond to the ear frequency response characteristics. (See Section 13.2.)

17.4 LOUDNESS OF A SOUND

Loudness of a sound is the magnitude of the auditory sensation produced by the sound. The units on the scale of loudness should agree with common experience in the estimates made upon the sensation magnitude. A true loudness scale must be constructed so that when the units are doubled the sensation will be doubled and when the scale is trebled the sensation will be trebled, etc. The sone is the unit of loudness. By definition a pure tone of 1000 hertz, 40 dB above a listener's threshold produces a loudness of 1 sone. The loudness of another sound that is judged by the listener to be n times the loudness of 1 sone is n sones. The loudness level of a sound, in phons, is numerically equal to the sound pressure level, in dB, relative to the threshold of 0.0002 microbar, of a free progressive plane sound wave of 1000 hertz which is judged to be equally loud.

The relation between loudness and loudness level is given by:

$$S = 2^{(P-40)/10} \qquad (17.1)$$

where
$$S = \text{loudness, in sones, and}$$
$$P = \text{loudness level, in phons.}$$

The loudness level of sones of other frequencies can be obtained from Equation (17.1) and Fig. 17.4 (phon levels and equal loudness contours).

17.5 FREQUENCY RANGE PREFERENCE FOR LIVE SPEECH AND MUSIC

Prior to 1947, the general impression from frequency preference tests of reproduced sound and from the frequency range of commercial sound-reproducing equipment was that the average listener preferred a restricted frequency range.

In order to obtain a better understanding of the reason for the apparent preference for the restricted frequency range in reproduced sound a fundamental and now classic all-acoustic test of frequency range preference was carried out in 1947. This test was an all-acoustic; the sound was not reproduced. Therefore, there were no electroacoustic transducers in the form of microphones, amplifiers, modulators, transmitters, records, receivers, pickups, reproducers, demodulators, loudspeakers, etc., used in these tests. The frequency discrimination was accomplished by means of acoustical filters.

The general arrangement of the test is shown in Fig. 17.5. An acoustical filter is placed between the orchestra and the listeners and is arranged so that it can be turned in or out. The acoustical filter is composed of three sheets of perforated metal to form a two-section acoustical filter, as shown in the figure. The frequency response characteristic of the acoustical filter shown in Fig. 17.5 approximated good commercial radio or phonograph reproduction in the high frequency range at the time of the test in 1947. The acoustical filter is composed of 10 units, with each unit pivoted at the top

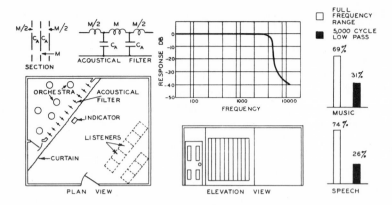

Fig. 17.5 Plan and elevation views of the arrangement of the apparatus for a frequency preference test for live speech and music. A sectional view, acoustical network and frequency response characteristic of the acoustical filter used in the tests are also depicted. In the acoustical filter: M = inertance of the series arm and C_A = acoustical capacitance of the shunt arm.

and bottom. The 10 units are coupled together and rotated by means of a lever. In this way the acoustical filters can be put in or out merely by turning the units through 90 degrees. The acoustical filters are shown in the full-frequency-range position in Fig. 17.5. A sheer cloth curtain which transmits sound with no appreciable attenuation over the frequency range up to 10,000 cycles and less than 2-dB attenuation from 10,000 to 15,000 cycles is placed between the acoustical filter and the listeners. The curtain is illuminated so that the listeners cannot see what transpires behind the curtain. The particular condition, that is, the full frequency range or 5000-cycle low-pass transmission, is shown on an AB indicator.

The tests were conducted in a small room which simulated an average living room in dimensions and acoustics. The orchestra was a six-piece dance band playing popular music. The average sound level in the room was about 70 dB. The changes from wide open to low pass to wide open, and so on, were made every 30 seconds. Two selections were played, and the listeners were asked to indicate a preference. The results of these tests, as shown in Fig. 17.5, indicated a preference for full frequency range. Similar tests have been made for speech. The preference in the case of speech is also for the full frequency range. There is a distinct lack of presence in speech with the limited frequency range.

The results of the tests described above were at variance with similar tests made upon reproduced sound prior to 1947. In 1947, three possible reasons were advanced for the variance between the live-sound and the reproduced-sound test results, as follows:

(1) The average listener, after years of listening to the radio and the phonograph, had become conditioned to a restricted frequency range and felt that this was the natural state of affairs.

(2) Musical instruments were not properly designed and would be more pleasing and acceptable if the production of fundamentals and overtones in the high frequency range were suppressed.

(3) The distortions and deviations from true reproduction of the original sound were less objectionable with a restricted frequency range. The distortions and deviations from true reproduction of the original sound fell into the following categories: frequency discrimination; nonlinear distortion; distortion due to spatial distribution by the loudspeakers, which varies with frequency; distortion due to a single-channel system, with no auditory perspective in the reproduced sound; phase distortion; transient distortion; distortion due to limited dynamic range; distortion due to differences in level of the original and the reproduced sound; distortion due to inadequate reverberation envelope; noise.

In more than the two decades that have passed since the frequency preference tests for live and reproduced speech and music were carried out, the reasons for the discrepancy between the two tests have been established as the distortions and deviations in the reproduced sound outlined in (3).

The results of the frequency preference experiment for live speech and music contributed materially to the initiation and advancement of wide frequency sound reproduction, that is, high fidelity sound reproduction.

17.6 FREQUENCY RANGE PREFERENCE FOR REPRODUCED SPEECH AND MUSIC

More than two decades have passed since the frequency range preference experiment for live speech and music was carried out. In the intervening time the advancements in sound reproduction have eliminated most of the differences between live and reproduced sound as outlined in (3) in Section 17.5. Therefore, it seemed appropriate to carry out another frequency range preference experiment for reproduced sound.

The experiment for determining the frequency preference for reproduced speech and music is shown in Fig. 17.6. The studio used in these tests is a small room of 7200 cubic feet with a reverberation time of 0.5 seconds. The sound is reproduced in a room of 3000 cubic feet and a reverberation time of 0.5 seconds. The acoustics and dimensions are practically the same as in the living room in a residence. The noise level in the studio is 16 dBA or NC6. The noise level in the listening room is 25 dBA or NC15. There are three sound-reproducing systems, namely, monophonic, stereophonic, and

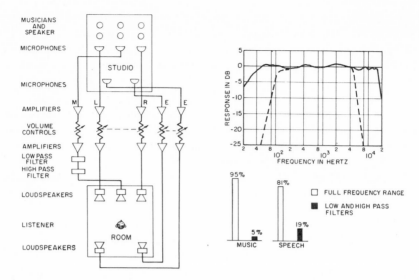

Fig. 17.6 Plan view of the apparatus for a frequency preference test for reproduced speech and music consisting of a studio, musicians and speakers, a monophonic channel *M*, stereophonic-bichannel *L* and *R*, reverberation envelope channels *E*, a listening room, and a listener. In the graph, the full frequency range is represented by the solid curve and the restricted frequency range by the dashed curve.

reverberation envelope. Each of the four channels consists of an RCA, 44BX Velocity Microphone, an RCA, BA31 amplifier, a special laboratory low-distortion power amplifier, and an RCA LC1C loudspeaker.

The overall frequency response characteristic is measured by supplying a constant input signal with respect to frequency to the chain consisting of the voltage amplifier, power amplifier, and loudspeaker. The sound output of the loudspeaker is measured with the same 44BX microphone. This method of response measurement provides an overall frequency response characteristic from sound input to the microphone to the sound output of the loudspeakers. The overall frequency response characteristics depicting the ratio of the sound-pressure output from the loudspeaker in free space to the sound pressure at the microphone in free space, with and without the electrical filters, are shown in Fig. 17.6.

The directivity pattern of the loudspeakers is very important when the listeners are located at relatively large angles with respect to the loudspeakers. In the loudspeakers used in these tests the variation in response at any frequency over a total angle of 90° is less that ±2 dB.

The nonlinear distortion is another important factor in reproduced sound. The overall nonlinear distortion is measured by supplying a distortionless signal to the input of the chain consisting of the voltage amplifier, power amplifier, and loudspeaker. The sound output of the loudspeaker is picked up by the microphone and fed to a harmonic analyzer. This method of measurement provides an overall distortion characteristic from sound input to the microphone to sound output of the loudspeaker. The total nonlinear distortion measured at the peak level of the reproduced sound is less than 0.25 percent. From the results reported below in Section 17.7, it will be seen that this value of nonlinear distortion is sufficiently low to be imperceptible.

An eight-piece orchestra was used in the tests. The change from full frequency range to restricted frequency range was made every 30 seconds.

For the reproduction of speech the monophonic reproducing system was used because speech is in fact a point source. There is no question but that there is a preponderance of preference for the full frequency range for the reproduction of speech as indicated in Fig. 17.6.

Two decades ago the frequency range preference tests for reproduced sound indicated a preference for a restricted frequency range. These tests employed monophonic sound reproduction. For music reproduction today, the sound reproduction can be carried out with auditory perspective and reverberation envelope. Therefore, to be meaningful with respect to the past, the reproduction of music was carried out with restricted frequency range for monophonic sound, and this was compared with full-frequency-range auditory perspective or stereophonic sound reproduction combined with full frequency range reverberation envelope. The latter system is termed quadraphonic sound reproduction. (See Section 6.6.) The results of the test depicted in Fig. 17.6 shows practically a unanimous preference for quadraphonic sound with the full frequency range as compared to a restricted-frequency-range monophonic sound.

17.7 SUBJECTIVE RESPONSE TO NONLINEAR DISTORTION IN REPRODUCED SOUND

The various elements in a perfect sound-reproducing system are absolutely invariant with respect to time and amplitude. However, in actual sound-reproducing systems in use today all the elements exhibit nonlinear characteristics to some degree. A nonlinear element introduces nonlinear distortion. The effect of nonlinear distortion is the production of harmonic components in the reproduced sound which are not present in the original sound. The introduction of spurious overtones leads to a deterioration of the reproduced sound. In order to obtain a better understanding of some of the effects of nonlinear distortion upon the reproduction of sound, subjective tests have been performed to determine the perceptible, tolerable, and objectionable nonlinear distortion in a sound-reproducing system as a function of the frequency range for different amounts of nonlinear distortion.

The effects of various types of nonlinear distortion upon the reproduction of speech and music have been determined by means of the sound-reproducing system shown in Fig. 17.7. The same studio and listening room are used as in the experiment in Section 17.6. The microphones, amplifiers, and loudspeakers are the same as those used in the frequency preference test of Section 17.6. The overall frequency response characteristic is shown in Fig. 17.6. The overall nonlinear distortion is less than 0.25 percent. The additional equipment consists of the distorting amplifiers, the complementary volume controls, and the variable low-pass filters. The overall distortion generated in the distorting amplifiers is a function of the level of the signal through the amplifier. By means of the volume controls 2 and 3 the level in the distorting amplifier and hence the distortion can be varied without varying the level of the reproduced sound because the attenuation or accentuation in volume control 2 is compensated by complementary accentuation or attenuation in volume control 3. Typical spectrums of the distortion for four different values of nonlinear distortion are shown in Fig. 17.7.

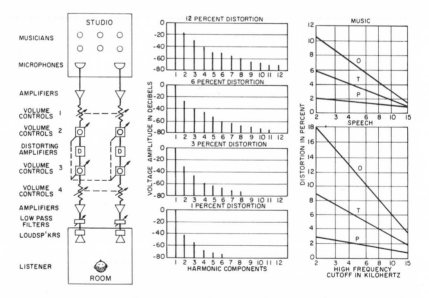

Fig. 17.7 Plan view of the apparatus for a subjective determination of the relation between nonlinear distortion and frequency range in reproduced sound consisting of a studio, musicians and speaker, a stereophonic two channel sound reproducing system with distorting amplifiers and low pass filters, a listening room and a listener. Typical spectrums for four different values of distortion are shown. In the graphs: Speech and music for different high frequency cutoffs. O = objectionable. T = tolerable. P = perceptible.

Since the same equipment was used as in Section 17.6, the overall nonlinear distortion with no additional nonlinear distortion introduced is 0.25 percent. All values of nonlinear distortion can be introduced from 0.25 percent to beyond 25 percent.

These tests were limited to three subjective gradations of nonlinear distortion, namely, perceptible, tolerable, and objectionable. Perceptible distortion is the amount of distortion in the distorting system required to be just discernible when compared with the reference system. Tolerable and objectionable are not as definite terms and are a matter of opinion. By tolerable distortion is meant the amount of distortion that could be allowed in low-grade commercial sound reproduction. By objectionable distortion is meant the amount of distortion that would be definitely unsatisfactory for the reproduction of sound in phonograph, radio, magnetic tape, and television systems.

Both speech and music were used in making these tests. In the case of music, an eight-piece orchestra was employed. The average results of a few of these tests, with a limited number of critical observers, are shown in Fig. 17.7. The amount of tolerable distortion is greater for speech than for music. Referring to Fig. 17.7, it will be seen that the amount of distortion which is judged perceptible, tolerable, or objectionable distortion decreases as the high frequency cutoff increases. This means that as the high frequency range is decreased, the amount of distortion which can be tolerated is increased. For example, these tests show that for music the tolerable nonlinear distortion in a sound-reproducing system with a 5000-cycle high frequency cutoff is of the order of 4 percent, whereas the tolerable nonlinear distortion in a sound-reproducing system with a 15,000-cycle high frequency cutoff is of the order of 1 percent. These tests show that unless the nonlinear distortion in a wide-frequency-range system is kept to this relatively low value, it will not be acceptable. While it is a comparatively simple matter to design, and it is relatively inexpensive to build, a sound-reproducing system with 4 percent nonlinear distortion and a limited frequency range, painstaking designs coupled with expensive components are required to achieve the relatively low order of nonlinear distortion of 1 percent in a system of wide frequency range.

17.8 FREQUENCY RANGES OF SPEECH AND MUSIC

The frequency range of the normal ear, for otologically normal persons in the age group eighteen to twenty-five years, is 20–20,000 hertz. Most musical instruments cover a frequency range to and beyond 20,000 hertz. However, the frequency ranges required for the reproduction of speech and music without any noticeable frequency discrimination by persons with normal hearing does not extend beyond 15,000 hertz. Specifically, the frequency ranges required to reproduce speech, musical instruments, and noises without any noticeable frequency discrimination is shown in Fig. 17.8. The reproduction of speech with subjectively perfect fidelity requires a frequency range of 100–8000 hertz and a volume range of 40 dB. The reproduction of orchestral music with subjectively perfect fidelity requires a frequency range of 30–15,000 hertz and volume range of 70 dB.

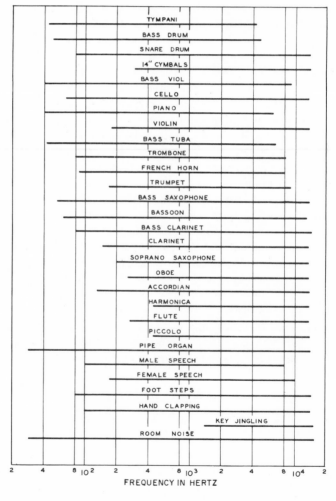

Fig. 17.8 The frequency ranges for the reproduction of speech, musical instruments and noises without any noticeable frequency discrimination. After Snow.

17.9 FREQUENCY RANGES OF SOUND-REPRODUCING SYSTEMS

The frequency ranges of the most common sound-reproducing systems are shown in Fig. 17.9. The frequency ranges shown are averages of existing systems. In specific cases the frequency ranges may be greater or less than those shown in the figure.

The frequency ranges of telephones vary over wide limits, depending upon the type of instrument, the central offices, and the interconnecting lines. The frequency range depicted is for instruments made in the last decade. Extending the frequency range would probably result in reduced articulation because of ambient room noise and noises produced by electrical interferences.

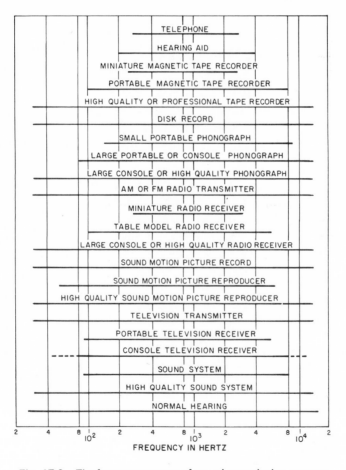

Fig. 17.9 The frequency ranges of sound-reproducing systems.

The frequency range of the hearing aid shown in Fig. 17.9 represents the average response of high-quality transistor hearing aids in use today. The low frequency range may be somewhat greater but in general, this added range cannot be used because of "rumble" and other low frequency noises.

The frequency ranges of magnetic tape recorders and reproducers are also shown in Fig. 17.9. The low tape speeds used in miniature and portable tape recorders limits the high frequency range. The small size of the loud-speaker and cabinet limits the low frequency response of the miniature and portable magnetic tape recorders. The frequency range of high-quality and professional tape recorders covers the entire audio frequency range. The commercial disc record also covers the entire audio frequency range, as shown in the figure.

The frequency ranges of phonographs are shown in Fig. 17.9. The size of the loudspeaker and cabinet determines the low frequency response of the portable and small console phonographs. The high frequency range is limited to provide proper balance. The high-quality phonograph covers the entire audio frequency range.

AM and FM radio transmitters also cover the entire audio frequency range. The frequency ranges of radio receivers are shown in Fig. 17.9. As in all sound-reproducing systems, the low frequency range is determined by the size of the loudspeaker and cabinet. The frequency ranges increase with the size of the radio receiver, as is shown in the figure. The high-quality AM receiver covers the entire frequency range if there is no cochannel interference; the high-quality FM receiver covers the entire audio frequency range.

The frequency range of a commercial sound-film record covers the entire audio frequency range, as shown in Fig. 17.9. The frequency range of a sound-motion-picture system, also in the figure, refers to the average of systems used in theaters. A small number of high-quality sound-motion-picture systems are in use, with frequency ranges varying from that of the commercial system up to that of the high-quality system shown in the figure. The frequency ranges of 16-millimeter and 8-millimeter film record and reproducers are restricted, compared to the commercial 35-millimeter film record and reproducers.

The frequency range of a television sound transmitter is also shown in Fig. 17.9. In general, the low frequency range is limited in transmission because of the high-ambient, low frequency noise level in the studio. The high frequency range is also limited by electrical noise and the frequency range of the audio-transmission networks. The low frequency range of the table-model television receiver is limited by the size of the loudspeaker and the cabinet. The frequency range of the large console television receiver is limited by the signal input.

The frequency range of the common low-cost sound system also appears in Fig. 17.9. For theaters, auditoriums, and other high-quality installations the frequency range of the sound system covers the entire audio frequency range. In some cases, the frequency range must be tailored to compensate for the poor acoustics of the auditorium or theater.

The hearing frequency range of a person with normal hearing (young ears) is also shown in the figure.

An examination of the characteristics noted in Fig. 17.9 shows that the consumer can obtain radio receivers, magnetic tape recorders and reproducers, and disc records and phonographs that cover the entire audio frequency range without discrimination. The data of the figure present a capsule representation of the frequency ranges of sound-reproducing systems. For more detailed information the reader should refer to Chapters 7–11 of this text.

17.10 AMPLITUDES AND SPECTRA OF SPEECH, MUSICAL INSTRUMENTS, AND ORCHESTRAS

The peak amplitude levels of speech, musical instruments, and orchestras are of value and interest in the reproduction of sound. The ratios of the peak pressures in discrete frequency bands, as shown by the discontinuities in the characteristics, to the average pressure for the entire frequency range are shown in Fig. 17.10. Except for the pipe organ and the bass drum, the peak levels are confined to the mid-audio frequency range.

17.11 EFFECT OF FREQUENCY DISCRIMINATION UPON THE ARTICULATION OF SPEECH

The frequency range required to reproduce speech without any noticeable frequency discrimination was described in Section 17.8. However, the frequency range for the intelligible transmission of speech is much less than that shown in Fig. 17.8.

The recognition or intelligibility of speech transmission is an important aspect of a sound-reproducing system. The sound-transmission system may be the air between the mouth of the speaker and the ear of the listener in the open air or a room, or it may be a sound-reproducing system such as the hearing aid, telephone, phonograph, magnetic tape reproducer, radio, sound reenforcer, sound motion picture, or television. In measuring speech recognition through a transmission system the speaker reads aloud speech

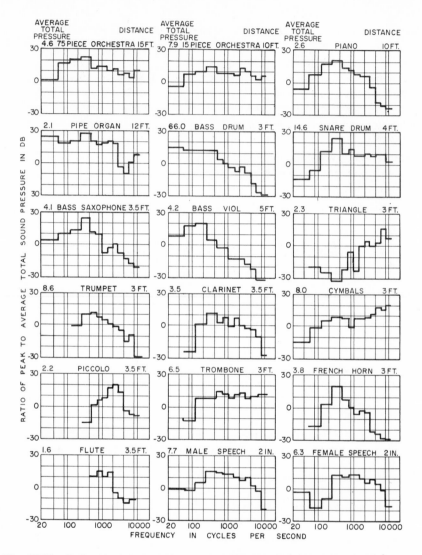

Fig. 17.10 Ratio of peak pressures in the frequency bands indicated to the total pressure in the entire spectrum for speech, various musical instruments and orchestras. The distances and the average total pressure in microbars are shown above each graph. After Sivian, Dunn, and White.

sounds, syllables, or words to a listener who writes down what he thinks he hears. A comparison of sounds, syllables, or words recorded by the listener with those uttered by the speaker provides the fraction that is interpreted correctly. The fraction is termed sound and syllable articulation and word and sentence intelligibility.

There are three types of articulation measurements involving sounds, syllables, and words. Sound articulation refers to the use of speech sounds such as "p," "a," "t," etc. Syllable articulation refers to the use of syllables such as "pat," "run," "eat," etc. Word intelligibility refers to the use of the complete word.

The effect of reducing the high and low frequency ranges upon the syllable articulation of speech at a normal conversational level is shown in Fig. 17.11. A consideration of Fig. 17.11 shows that a relatively high articulation can be obtained with a very narrow transmission band. However, the quality of the reproduced speech is very much impaired by transmission over a narrow frequency band. From the standpoint of articulation, a limited frequency range may be actually superior to a wider frequency band, because of the introduction of additional noises and distortions in a wider band, unless particular precautions are observed. In the case of speeches, plays, and songs, a limited frequency range impairs the quality and artistic value of the re-produced sound.

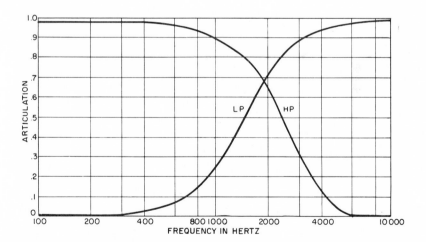

Fig. 17.11   The effect of frequency range upon syllable articulation of speech. *HP* = The use of a high pass filter with all frequencies below the frequency given by the abscissa of the *HP* curve removed. *LP* = the use of a low pass filter with all frequencies above the frequency given by the abscissa of the *LP* curve removed.

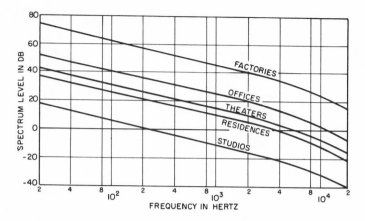

Fig. 17.12   The average noise spectrum for factories, offices, theaters, residences and studio. The curves show the sound pressure level for one hertz bandwidth. 0 dB = 0.0002 microbar.

17.12 ROOM NOISE AND THE REPRODUCTION OF SOUND

The threshold hearing characteristic of Fig. 17.4 sets the lower limit for an ideal transmission system with the listener in a very quiet place. The ideal of the ambient noise being below the threshold of hearing is never realized by listeners under practical conditions because the lower limit of hearing is determined by the ambient room noise.

The average noise spectrum characteristics for factories, offices, theaters, residences, and studios are shown in Fig. 17.12. The data in the figure are for average conditions; that is, 50 percent of the locations exhibit a higher noise level and 50 percent of the locations exhibit a lower noise level. The noise spectrum characteristics are practically all of the same shape, as shown in Fig. 17.12.

The lower level of hearing limits for pure tones can be derived from the spectrum characteristics of Fig. 17.12 and are depicted in Fig. 17.13. A pure tone for each of the conditions cannot be heard below the characteristics of Fig. 17.13. For example, in the mid-frequency range in the average residence a pure tone cannot be heard below 25 dB. If sound is reproduced at a top sound level of 75 dB, this provides a volume range of 50 dB, which is quite adequate for high-quality sound reproduction.

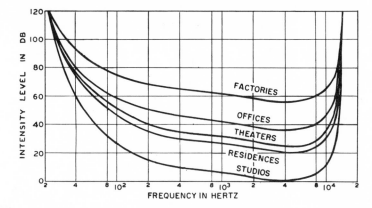

Fig. 17.13   Hearing limits for pure tones. The characteristics are for a typical listener in a typical factory, office, theater, residence and studio. Pure tones with levels below the curves cannot be heard.

17.13 HEARING DAMAGE DUE TO HIGH NOISE AND SOUND LEVELS

Persons exposed to high sound and noise levels beyond 85 dBA may suffer damage of the human hearing mechanism.

The United States Government has established guide lines on noise and sound levels as follows:*

(1) Protection against the effects of noise exposure shall be provided when the sound levels exceed those shown in Table 17.1 of this section when measured on the A scale of a standard sound level meter at slow response.

(2) When employees are subjected to sound levels exceeding those listed in Table 17.1, feasible administrative or engineering controls shall be utilized. If such controls fail to reduce sound levels to within the levels of the table, personal protective equipment shall be provided and used to reduce sound levels to within the levels of the table.

(3) If the variations in noise level involve maxima at intervals of 1 second or less, the noise is to be considered intermittent. In such cases, when the duration of the maxima are less than 1 second, they shall be treated as of 1-second duration.

*Article 50-204.10, "Occupational Noise Exposure," Federal Register, Vol. 34, No. 96, Tuesday, May 20, 1969. Section 1910.95, 10518. Occupational Safety and Health Standards, Federal Register Vol. 36, No. 105, Saturday, May 29, 1971.

TABLE 17.1

Permissible Noise Exposures

Duration Per Day in Hours	Sound Level in dBA
8	90
6	92
4	95
3	97
2	100
$1\frac{1}{2}$	102
1	105
$\frac{1}{2}$	110
$\frac{1}{4}$ or less	115

When the daily noise exposure is composed of two or more periods of noise exposure of different levels, their combined effect should be considered, rather than the individual effect of each. If the sum of the following fractions: $C_1/T_1 + C_2/T_2 \cdots C_n/T_n$ exceeds unity, then, the mixed exposure should be considered to exceed the limit value. The term C_n indicates the total time of exposure at a specified noise level, and T_n indicates the total time of exposure permitted at that level.

(4) In all cases where the sound levels exceed the values shown in the table, a continuing, effective hearing conservation program shall be administered.

Exposure to impulsive or impact noise should not exceed 140 dBA peak sound pressure level.

The main source of noise is machinery of all kinds. Therefore, every effort should be made to select and design quieter machinery. Designing-out machinery noise is more practical than getting rid of it after the equipment has been built. There are still other high sound levels that present a danger to hearing, namely, the high sound levels of modern rock music. The average sound levels of some bands exceed the damage risk criteria. In this connection, the levels of sound in monitoring rooms exceed the danger sound levels. Many young people reproduce sound from radio, disk records, and pre-recorded magnetic tapes at sound levels beyond those shown in Table 17.1. If noise control through engineering cannot reduce noise exposure to safe levels protective devices must be used in the form of ear plugs or ear muffs. These devices will provide 20 to 30 dB attenuation over the main part of the audio frequency range.

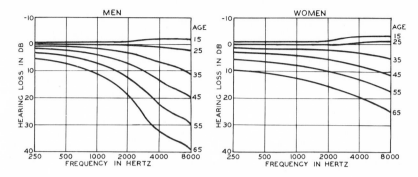

Fig. 17.14   Hearing loss frequency characteristics for men and women of various ages. The characteristics are averaged from several different investigations.

17.14 HEARING ACUITY

The hearing response with respect to sound level and frequency is an important aspect of sound reproduction. Hearing loss for men and women as a function of the frequency is shown in Fig. 17.14. These data show that the general overall hearing acuity decreases with age. In particular, the hearing loss is most pronounced in the high frequency range. Listening to reproduced sound results in an upset of the balance of high and low frequency sounds. There is also a loss in the perception of the high partials or overtones. The musical talents which are least affected by a hearing loss are pitch, duration, time, vibrato, rhythm, and to some extent loudness. Those most affected are timbre, growth, decay, volume, and presence.

17.15 SUBJECTIVE ASPECTS OF SOUND REPRODUCTION

Home entertainment in the audio complex is the reproduction of sound. Sound reproduction is the process of picking up sound at one point and reproducing it either at the same point or at some other point, either at the same time or at some subsequent time. The general objective of sound reproduction, as the term implies, is to provide a resemblance to the original sound, but usually modified by certain subjective aspects as directed by the desires of the listener. Up until very recent times the objective was to provide a perfect transfer characteristic between the input and the output of the sound-reproducing system. In a perfect transfer characteristic there are constant relationships between the fundamental input and output parameters that define the signal. In the ideal transfer characteristic the relationship between the fundamental input and output parameters is modified as dictated by subjective aspects involving realism, sensationalism, and emotionalism. In general, in order to attain the desired ideal transfer characteristic by the application and implementation of the appropriate subjective aspects of sound reproduction, one must start from a perfect transfer characteristic.

A. Physiological and Psychological Factors in Sound Reproduction

The physiological and psychological or subjective processes involved in listening to live or reproduced sound consists of two parts, namely, the sensational response and the emotional response. The sensational response is invariant from person to person, within the individual variations of sensorial perception from person to person. The emotional response varies depending upon the individual, the program, the rendition, and the era involved. For example, popular music of any era has great appeal to the younger generation of that particular time. On the other hand, the popular music of the 20's and 30's appeals to an older generation but has very little interest for the teen-agers of today.

1. Sensational Process. The sensational response is the subjective response involving the sensory performance of the human hearing mechanism. The main physical factors involved in the sensational response in the reproduction of sound are frequency range, nonlinear distortion, loudness, spatial effects, and reverberation.

The sensational response of the listeners may be expressed as:

$$R_{sn} = f_n(p_n) \qquad (17.2)$$

where $R_{sn} = R_{s1}$, R_{s2}, etc., are the sensational responses to the particular parameters $p_n = p_1, p_2$, etc., of the transmitting system. In this consideration the parameters in the transmitting system are as follows: p_1 = frequency range; p_2 = nonlinear distortion; p_3 = loudness; p_4 = auditory perspective; p_5 = reverberation.

Referring to Equation (17.2), the highest sensational response is obtained under the following conditions: full audio frequency range, low or imperceptible nonlinear distortion, a certain desirable and tolerable peak level of sound reproduction depending upon the size of the room, true auditory perspective, and an optimum reverberation time.

Subjective responses to the above physical factors are quite definite; consequently, variation in sensational response from person to person is quite small indeed.

2. Emotional Response. The emotional response is very complex because there is indeed a very wide variation in the emotional response to speech and music. The properties of a musical tone are frequency (pitch), intensity (loudness), growth, duration, decay, portamento, timbre, vibrato, and deviations. Music may be defined as the art of producing a series of pleasing, expressive, intelligible, or sensational tones or combinations of tone complexes. The performer employs the properties of a tone and the arrangement and timing of the tone to create the desired emotional response for the audience. The establishment of the desired emotional response is not amenable to the scientific method, unlike the case for the sensational response. Those involved in the production of emotional responses employ for the most part empirical and intuitive means.

The emotional responses of the listener may be expressed as:

$$R_{en} = {}_n(L_n, S_n, Q_n, A_n) \qquad (17.3)$$

where $R_{en} = R_{e1}, R_{e2}$, etc., are the emotional responses to the particular factors L_1, S_1, Q_1, A_1; L_2, S_2, Q_2, A_2; etc., and L_n = listener n, S_n = reproduced program n, Q_n = reproduced rendition n, and A_n = the particular era n. As stated by Equation (17.3), the emotional response depends upon the particular listener, the rendition, the reproduced program, and the era. Under these conditions, there is to be expected a wide variation in response, depending upon the factors L, S, Q, and A.

Even in the case of classical music where the era extends over several decades, there is a wide variation in response among listeners to a particular program depending on the manner of the rendition. In the case of popular music there is a high degree of variation in emotional response involving the era, the reproduced program, the rendition, and most of all the listener.

B. Total Subjective Response

The total subjective response (combined response due the physiological and psychological processes) is a function of the elementary sensational response and the emotional response:

$$R_{tn} = F(R_{sn}, R_{en}) \qquad (17.4)$$

where R_{tn} is total subjective response, R_{sn} is given by Equation (17.2), and R_{en} is given by Equation (17.3).

If the physical characteristics of the transmitting system exhibit poor qualities, then the sensational response R_{sn} will be relatively poor. Under these conditions, even though the emotional response may be good, the total subjective response will be poor. Or, good subjective response cannot be expected if physical performance characteristics are poor.

REFERENCES

Bauer, Benjamin B. "Octave-Band Spectral Distribution of Recorded Music," *Journal of the Audio Engineering Society*, Vol. 18, No. 2, p. 165, 1970.

Dunn, H. K. and White, S. D. "Statistical Measurements on Conversational Speech," *Journal of the Acoustical Society of America*, Vol. 11, No. 3, p. 278, 1940.

Fletcher, Harvey. *Speech, Hearing, and Communication*, Van Nostrand Reinhold, New York, 1953.

French, N. R. and Steinberg, J. C. "Factors Involving the Intelligibility of Speech Sounds," *Journal of the Acoustical Society of America*, Vol. 19, No. 1, p. 90, 1947.

Hoth, D. F. "Room Noise Spectra at Subscriber's Telephone Locations," *Journal of the Acoustical Society of America*, Vol. 12, No. 4, p. 499, 1941.

Miyagawa, Rikuo, Nakayama, Takeshi, and Miura, Tanetoshi, "Design of Reproduced Sound Quality by ESP Method," *Reports of The 6th International Congress on Acoustics*, Tokyo, 1968.

Normal Equal-Loudness Contours For Pure Tones and Normal Threshold of Hearing Under Free Field Listening Conditions, International Organization for Standardization, ISO Recommendation R226, 1961.

Olson, Harry F. *Acoustical Engineering*, Van Nostrand Reinhold, New York, 1957.

Olson, Harry F. *Music, Physics, and Engineering*, Dover Publications, New York, 1967.

Sivian, L. J., Dunn, H. K., and White, S. D. "Absolute Amplitudes and Spectra of Certain Musical Instruments," *Journal of the Acoustical Society of America*, Vol. 2, No. 3, p. 330, 1931.

Index

329